LOCAL ENVIRONMENTAL CHANGE AND SOCIETY IN AFRICA

Local Environmental Change and Society in Africa

2nd Edition

Edited by

M.A. MOHAMED SALIH
Professor of Politics of Development,
Institute of Social Studies,
The Hague, The Netherlands
& Department of Political Science,
University of Leiden, The Netherlands

SPRINGER-SCIENCE+BUSINESS MEDIA, B.V.

A C.I.P. Catalogue record for this book is available from the Library of Congress.

ISBN 978-94-010-3878-2 ISBN 978-94-010-1003-0 (eBook)
DOI 10.1007/978-94-010-1003-0

First published 2000
Second edition 2001

Contents

Preface

The Organization for Social Science Research in Eastern and Southern Africa (OSSREA) is a network of social scientists in the sub-region established in 1980. Its main objectives are to encourage and promote interest; to engage in collaborative research; to facilitate scholarly exchange of ideas and publication between individuals and institutions engaged in the study and research in the social sciences. OSSREA also aims to promote the training of social scientists from the sub-region in the study of and research in the social sciences and to encourage establishment of institutions dedicated to this goal. In pursuance of and consistent with its objectives, OSSREA established a fund to provide research grants and training fellowships to social scientists from the sub-region.

Every year OSSREA offers, on a competitive basis, two limited types of research grants: the first on topics related to the social sciences in general and the second on gender issues in particular. The grants assist the social scientists to conduct field research in an eastern or southern African country, society, institutions, sector, or problem area. The objectives of the social science competition are intended to encourage and promote upcoming scholars from the sub-region to conduct research in the social, economic, political or other problems. Moreover, it aims to strengthen and raise the capacity of faculty members and students in field research, and promote the subsequent availability of research findings for teaching as well as to inform individuals and institutions.

The second research competition focuses on gender issues and is intended to encourage local scholars to conduct relevant research that is expected to contribute to the education and understanding of gender issues and its integration into development thinking, planning and implementation. Another objective is to avail the research findings to policy makers as well as for teaching. Themes such as ethnic conflicts, analysis, management and resolution; regional co-operation; sustainable development and natural resource management, among others, and their relationship to gender; development; governance have been researched by our grant winners. OSSREA's research grant fund aims towards building up the capacity of eastern and southern African research institutions by offering an opportunity to up-coming researchers and scholars to address topics that are important to their work domain as well in informing as decision-takers and policy-makers. A jury of distinguished scholars from the sub-region is selected every year to evaluate the grant applications. The grant winners are brought together for a research methodology workshop, where they respond to comments and questions raised by the jury and other resource persons. It is at this point that the grant winners sharpen their research focus and exchange experiences with each other, the jury and the resource persons before commencing the fieldwork. Over the past 16 years some 300 research reports have been completed. Some of these reports have been published in OSSREA Research Report Series. Others have been summarized and published as articles in OSSREA's bi-annual journal: The Eastern African Social Science Research Review (EASSRR), now in its 17th year of existence. EASSRR is widely distributed within and outside the sub-region.

This book is the first attempt to edit some recent research material dealing specifically with local environmental change. By focusing on local environmental change and society, the reader is offered an opportunity to obtain a comparative perspective on topics pushed aside by an apparent interest in global environmental issues. It has been a deliberate decision by the researchers from the sub-region to bridge the knowledge gap in areas vital to the survival of millions of peoples. It is also an acknowledgement that the researchers feel a moral and ethical responsibility towards society by producing usable knowledge that could assist to those engaged in development interventions and to make them aware of the environmental cost of unchecked interventions. OSSREA hopes that similarly edited volumes on other topics studied by researchers reflecting on the realities of the sub-region will appear in future volumes.

OSSREA's research, training and publications, including this volume could not have been possible without the generous support of a number of donor organizations. OSSREA extends its acknowledgement and gratitude for the generous support of the organizations and foundations that have been instrumental in assisting it to achieve its objectives. The research grants through which this work has been accomplished deserve special mention and applaud. Among the major organizations and foundations that have supported OSSREA are: The Ford Foundation, the International Development Research Center (IDRC), the Norwegian Agency for Development Co-operation (NORAD), the Swedish International Development Cooperation (Sida/SAREC) and The Netherlands Directorate of International Development (DGIS). Their invaluable and indispensable support has allowed OSSREA to make significant impact on the development of the social sciences in the sub-region and beyond. It is hoped that the reader will find this volume academically inspiring, interesting and useful.

Professor Abdel Ghaffar Mohamed Ahmed
Executive Secretary
Organization for Social Science Research
In Eastern and Southern Africa (OSSREA
Addis Ababa, Ethiopia

28 June 2001

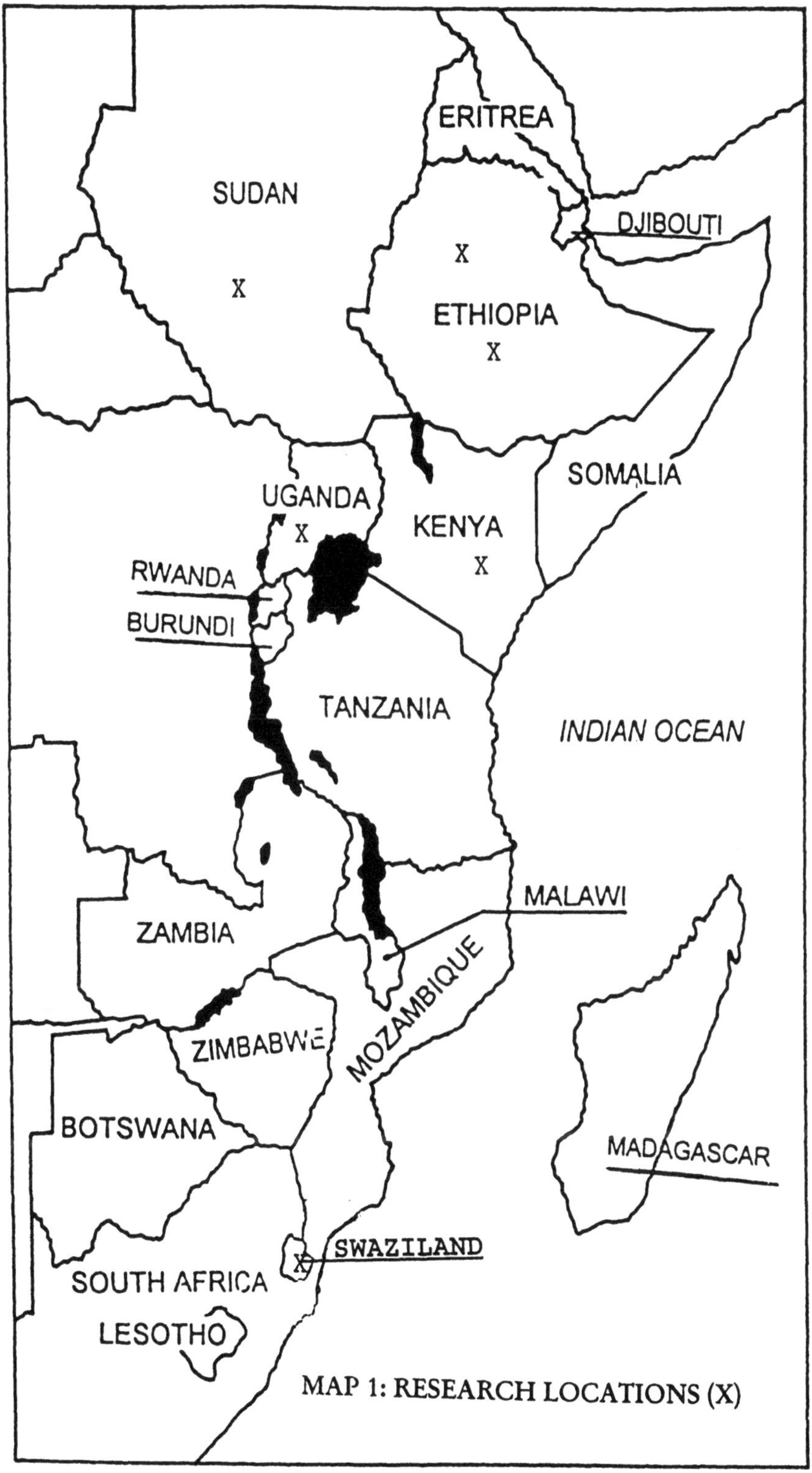

MAP 1: RESEARCH LOCATIONS (X)

List of Tables

1 Introduction: Elements of Local Environmental Change

M. A. Mohamed Salih

During the course of African history, complex economic and technological developments have unleashed unprecedented social and environmental transformations. From pre-colonial, colonial to post-colonial history the African continent has been the subject of economic expansion, industrialisation and urbanisation that has produced momentous social transformations. These transformations are neither neutral nor without social costs and benefits and as such have produced many a loser or gainer. No wonder, environmental studies have been concerned largely with calls about environmental marginalisation and the appropriation of local natural resources by private or state interests (Kousis 1998, Mohamed Salih 1999). A generally accepted conclusion is that environmental alienation has contributed to the distortion of local ecosystems and economies, thus denying the local communities the right to eke out their living from a productive and healthy environment. This theme has attracted a large number of scholars who have held the front, so to speak, against ecological marginalisation and its manifestation in various forms of social and economic marginalisation.

However, this volume aims at articulating a local level understanding of the social and economic factors that have contrived to shape the twin processes of environmental change and social transformation in four African countries (Ethiopia, Kenya, Uganda and Swaziland). It presents the results of the research carried out by four African social and natural scientists dealing with various aspects of environmental change and its consequences on society. The main objectives of the research from which this volume emanates are fourfold: First, to provide detailed case studies to explain the relationship between social and environmental transformation and their consequences on the African conditions. Second, to explore the extent to which an impaired environment function can also contribute to ruptures in the social and economic conditions of society and vice versa. Third, to examine local responses to major environmental problems, bearing in mind the common pessimism that conflates environmental change and environmental problems. Fourth, to explicate the implications of the case studies to expanding concern with public policy dialogue involving the stakeholders affected by local environmental change.

Instead of treating local communities as passive victims of imposed environmental change, the case studies presented in this volume assess how local communities have, despite many difficulties and external pressures, been able to adapt their resource management systems in response to environmental change. However, in most cases the pressures have become so high and complex that they

M.A.M. Salih (ed.), Local Environmental Change and Society in Africa, 2nd Edition, 1–16.

have not only generated problems never before experienced by these communities, but also that the communities do not possess the knowledge necessary to mitigate the consequences. The problem is the often-neglected local dimension of environmental change, but one of impaired society-environment relations that have simultaneously had an impact on ecosystem and economic sustainability. However, familiar concerns with the non-sustainability of top-down policies imposed on local environmental managers have not yet gone away. Despite at least three decades of struggle for popular participation in local resource management, global conventions and national policy pledges have not been matched by practice and words are still louder than action.

Although the seven case studies presented here do not capture the totality of socio-economic and environmental conditions prevalent in the African continent, they invariably reveal that little progress has been made in translating public concerns with the environment into effective tools for socio-economic or environmental sustainability. The technology used is invariably inappropriate: too rudimentary (hoes, axes, machetes) to produce food self-sufficiency at the household level, with high and no built-in safety mechanisms to mitigate social cost (irrigation) or absent in the case of pollution control.

The case studies echo Joan Davidson's (1991:143) remark that, "a major problem is apathy – and the generation of words rather than action". Without understanding the pattern and magnitude of local environmental change on income, nutrition, human health, water, sanitation, gender, self-esteem and other things through fieldwork based research it will be inappropriate to devise policy interventions based on some social scientists' imaginative capacities. In some of these circumstances, 'research tourism' has created more confusion than answers. Furthermore, the global environmental ethics that have generated tens of thousands of publications and culminated in thousands of environmental conferences, conventions, treaties, laws and socio-economic instruments are yet to be translated into workable local level environmental interventions.

LOCAL ENVIRONMENTAL CHANGE

The notion of environmental change used in this volume denotes a variety of processes that have resulted from human-intervention in the physical environment. Environment here is widely defined to encompass our surroundings, including nature, natural resources and the socio-economic and physical infrastructure created by human beings. The emphasis on a wider definition of the environment stems from the variety of environmental functions (production, sink, regulatory, information carriers/education and aesthetic-symbolic among other functions) and their relationship to the production and reproduction of society. Environmental change results from the intensity, scale and quality of human interventions in the environment, through the expansion of agriculture and livestock production, irrigation, deforestation, industrialisation and urbanisation to mention but a few. In other words, the excessive use of environmental functions beyond their carrying capacity may in some circumstances transform environmental change into serious environmental problems.

Tallegen and Wolsink (1998:1-15) classify the common environmental problems into three categories: First, exhaustion or depletion of renewable and non-renewable resources. The work of Meadows et al (1972 and 1992) among others falls within this category. Second, pollution or the transformation of matter or substance into solid, fluid or gaseous waste, which is then released into the environment and can be a cause of environmental change as well as inducing environmental problems. Climate change, ozone depletion and greenhouse effects are among the common consequences of pollution. The environmental problems resulting from excessive use of the sink and regulatory functions of the environment are related to the environment's incapacity to handle excessive pollution and waste absorption. Third, disruption of the environmental functions, either by natural or human induced interventions as a major source of environmental change. For instance, the unintended consequences of fertilisation in its capacity to generate organic substance for natural regeneration or the consequences of reducing fresh water flow on marine life. Nature-society relations have been succinctly summarised by Martell (1994:24) who argues that, "the society-nature relationship is constituted by natural limits on society, society's effects on nature and the effects of society's impact on nature as they rebound on society". To illustrate this point, I quote Goldblatt (1994: 29) at length,

> An abstract model of the relationship between societies and environment would differentiate between direct causes of environmental degradation and its structural causes. It would differentiate between the capacities and potentialities of different systems of production as agencies of environmental change, and combine this with a sensitive appreciation of the impact of demographic change. It would trace the causes of social actions that affect ecosystems, particularly economic and demographic behaviour, though it would not restrict the causes of that change to economics and demographics alone. It would specify the type of material impacts produced, known and unknown, and their multiple ecological and social consequences. It would be sensitive to the range of cultural perceptions, knowledge and valuations of the environment. It would trace the implications of ecological changes, distinguishing between those perceived by participants and those operating behind their backs, between those affecting economics, demographics and health and those contravening moral-aesthetic standards. It would do so with reference to the sorts of demands and requirements that different systems of production place on the environment.

If negative environmental change impairs the environmental functions necessary for the maintenance of nature and the natural resources on which society depends for production and reproduction, its consequences must be social. However, these consequences are not neutral as we may think. Adams (1992: 87) reminds us that, "the environment is not neutral in its effects on the poor; environmental quality is mediated by society, and society is not undifferentiated. Access to and the distribution of environmental 'goods' (be they cultivable land, fuelwood or clean air) is uneven". In agrarian-dependent societies, such as is the case of Africa, the relationship between society and environment militate an association between socio-economic and environmental well being. While society's unchecked intervention on the environment may impair its capacity to eke out a living from a healthy environment, what Chambers (1988) calls the "physical environment of poverty" is in itself not conducive to the production and reproduction of liveable socio-economic conditions. To that extent, although game parks, rainforests and the wilderness are important, for agrarian Africa, environment constitutes more than the beauty of its benign forms. The environment constitutes first and foremost the socio-economic and physical conditions or the surroundings from which people eke out a living.

Recent reports on social and economic conditions reveal the interconnectedness between environmental and socio-economic well being. The economic crisis that had besieged the continent from the 1970s to the 1990s show that the human health environment has deteriorated because of the deterioration of the economy. The linkages between human environmental health and economic "health" have become even more revealing when the health situation is examined. It has been reported by the African Development Bank (1998), that in the health sector, 20 medium and low spending countries decreased their per capita expenditure on health on average by nearly 10 percent. An earlier picture, which emerged from Basu and Stewart (1995: 157-9) research reveals that, "on average, countries spent less on health during adjustment years than non-adjustment years, but no statistically significant relationships between health expenditure and adjustment were identified". Curbing allocations for the social sector and the reduction of subsidies for health, food, education and extension were among the main credo of structural adjustment policies and their neo-liberal philosophy.

According to UNICEF (1994), the number of people per physician worsened in over half of countries. The same applies to the ration of nurses to population, which also deteriorated in one-third of the African countries. Deteriorating real incomes due to high inflation and substantial reduction of subsidies and rising charges had adversely affected the poor ability to use the already debilitated health services. Africa Development Report (1998:161) summarises the problem as follows:

> In many African countries, systems of health care are failing to provide services that can make the greatest impact on major causes of illness and disease... Poorly trained personnel and insufficient supplies and drugs at the primary care level prompt residents to seek primary health care from intermediate level facilities and hospitals. Efforts to reform the system in many countries suffer from organisational problems. In addition to organisational problems, health services in the poorest countries also suffer from inadequate financial support.

Inadequate financial support and decrease in social allocation ratio, as I have mentioned earlier, is a direct result of structural adjustment to the extent that many health problems, which Africans thought had gone, are returning, excluding AIDS and other epidemics. For instance, malaria is estimated to represent 2.3 per cent of the overall global disease burden. Malaria represents 9 per cent of the disease burden in Africa, where it remains one of the leading causes of morbidity According to the Africa Development Report (1998:154), an estimated 300-500 million cases of malaria are reported every year, each year resulting in about one million deaths of mainly children under five. Tuberculosis (TB) has become the leading infectious disease; with one and a half million case a year it causes almost 600,000 deaths per year. The so-called TB/HIV co-epidemic represents one third of the new cases. WHO (1997:34) reports that, despite its eminent threat to Africa, TB remains a low priority among governments and international development agencies? Health concerns cannot be separated from concerns with education and food security. This in view of this author is a major global policy setback in the area of human health where diseases such as TB/HIV know no boarders. In essence it is the local communities with failing health facilities that suffer from the human environmental health problems which emanate from this global failure.

Among the contributors to human health problems is food insecurity and malnutrition due to human (policies, prices etc.) and environmental (drought, soil infertility etc.) factors that deprives people's entitlement to food. I argue elsewhere (Mohamed Salih 1994:10), that biased top-down food policies tend to forsake the interests of the poor and to deny them access to land and agricultural inputs. Generally, such policies deprive people of access to the productive assets necessary to produce sufficient food for them to be able to live a healthy and productive life. The book has also demonstrated that as a result of structural adjustment policies, (1) an increasing number of people can no longer meet the minimum food requirements for a healthy life. (2) The number of people living below the poverty line has increased, therefore increasing the incidence of malnutrition. (3) People's economic power has declined at the macro-level therefore reducing their ability to purchase and produce food, thus contributing further to food insecurity. The FAO conference on Food Security in Africa (1996) concludes its sombre report on the food situation in Africa as follows:

> Current estimates indicate that approximately 35 percent of the region's population or some 217 million people are currently chronically undernourished - 213 million are in low-income food deficit countries (LIFDCs).... The number, scale and intensity of emergencies in Africa have all been increasing due to both natural disasters (especially droughts) and human-caused calamities including civil strife and conflict. Wars and related factors have become the single most serious cause of food insecurity in much of the region. For example, in the countries covered by the Intergovernmental Authority on Drought and Development (IGADD) some 45 million people (nearly half of the population) were estimated to be subject to food insecurity in 1990. Of these, some 45 percent were classified as refugees and affected by war. In 1994, out of the world total of 32 million victims of disasters receiving relief assistance from the World Food programme (WFP), 21.5 million were living in Africa. Of these, nearly two-thirds were the victims of human-caused disasters, distributed among West and Central Africa (4.6 million), the Horn and East Africa (3.9 million) and Southern Africa (5.6 million).

The broad definition of environmental change used in this introduction maintains a strong presence in Africa, albeit with varying degrees of intensity, patterns and magnitude. What I have described as environmental problems at the physical environment level (exhaustion, pollution and disturbance) is a result of the long history of African and non-African exploitation of the environment as well as a marked population growth not accompanied by a similar pace of capital growth or technological development. This process has been enhanced by global process that enhance foster exclusion while caring less about community and locality. Meadows et al (1992:39) captured this point in relation to the developing countries in general:

> In poor African countries capital growth has a hard time keeping up with population growth for many reasons – because inevitable surplus is siphoned off to foreign investors, to the luxury of local elite, to debt repayments, or exorbitant militarization. Because there is too much poverty, technical inefficiency, or

> mismanagement to generate an investable surplus in the first place, he population is stuck in a pattern of growing bigger without growing richer.

It is not possible in the African context to isolate environmental from social problems, which explains why international development institutions have been concerned with the triple consequences of the slow pace of economic growth on the social and environmental conditions of the peoples of Africa.

Some local environmental changes have affected socio-economic sustainability and reduced the productivity of natural resources vital for the local communities as well as national economic development. For instance, wetland reclamation for agricultural production purposes has been a direct result of cultivable land scarcity due to inequitable distribution or excessive population growth. The case of Uganda shows that the economic gains from rice cultivation by small number of wealthy farmers have been give priority over benefits that were traditionally associated with them. However, Barbier's (1991: 44) view that, "formal property law and titling regulations often ensure that clearing of land is a prerequisite for guaranteeing claims to frontier land holdings", has been challenged in the case of Uganda. Private landowners who cultivate fire wood plantations to supply energy for urban consumption have deliberately taken up private reforestation as a profitable business. This view goes along with Shepherd's (1991:63) contention that,

> The problem has been, and remains, that it is not economic to grow fuel for the urban market in the place where the trees were being cut for that market. Once trees have gone from remote open areas, we begin to find that they are replaced by trees planted scores and even hundreds of miles away in farms much nearer to towns, so that the reduced transport costs of delivering this wood to urban markets offset the fact that tree-growing is more costly than felling pre-existing bush-land trees.

However, the Uganda case proves Shepherd's view that peri-urban plantations are uneconomic and wrong. In fact the Uganda case study shows that peri-urban tree plantations are profitable, yet their wood prices are too expensive for the urban poor to purchase. In fact the energy crisis in Uganda is such that poor households are increasingly indebted to a wood seller to the extent that a dependency relationship has emerged between the two.

Despite decades of research and development projects concerned with mitigating the consequences of land degradation and soil erosion, these two negative forms of local environmental change continue to pose major challenges to African agriculture and livestock producers. Although the chapters show that peasants continuously modify traditional land management practices, such modifications have been either very slow in coming or have been confronted by serious shortages of modern delivery input delivery systems. Technological change has often failed to provide desirable solutions and in some cases has created its own social problems that require urgent policy interventions. The human health problems associated with irrigation development in Kenya provide a classic case of how - although technological interventions have their own internal dynamics, they equally result in social costs that traverse their immediate impact area.

Local environmental change has transformed the nature of the household, status role relations and the traditional patterns of a sexual division of labour. On the one hand, traditional forms structures have provided security; on the other hand, new realities demand prudent transformations to cope with new demands and structural socio-economic and political changes. Processes of individualization have fundamentally changed the nature of the family life cycle. The tension between continuity and schism and the oscillation between tradition and modernity are inseparable from the environmental consciousness brought about by real or human-made resource scarcity-cum-degradation, economic pressures and social ruptures. The maintenance of a viable life is no longer a function of meeting socio-economic needs, but maintaining a sustainable environment that can generate both continuity and schism.

STRUCTURE OF THE BOOK

This volume consists of case studies based on field material collected from rural and urban areas between 1995 and 1997. A theme common to all case studies is that the research has been conducted with three question in mind: 1) what are the specific environmental pressures and on what resources (water, soil, forests, health, etc.), 2) what are the causes and consequences of these pressures on society, and 3) what national policy measures and local solutions have been adopted in order to eliminate, contain or minimize these pressures. Although the case studies have utilized a variety of methods and theoretical strands, the authors have commonly combined quantitative and qualitative methods taking advantage of their long experiences with the field situations they have investigated. Each case study has been concluded with a set of policy recommendations some of which are of a general nature and I have therefore incorporated them in the introduction. The chapters are also organized in such a way that those dealing with similar environmental issues are sequentially clustered so that the reader can move from one chapter to another without significant rupture.

Mwangi's chapter sets out to identify the causes of environmental and health problems in two schemes, evaluates the existing institutional arrangements at the household, scheme and national planning levels in dealing with these problems and finally gives a framework for irrigation development planning which integrates environmental and health considerations. The material presented is cross-sectional and exploratory in nature and reveals in detail the environmental and health problems in irrigation projects and the existing planning initiatives in irrigation. Problems like water logging, alkalinity and salinity were also found to be associated with irrigation development. As a result, environmental health problems have increased in the recent past, posing a great threat to human health and irrigation development. In explaining the environmental health threats posed by irrigated agriculture in Kenya, Mwangi argues that improper understanding of the complexity of irrigation agro-ecosystems and their interaction with a wide range of vectors have made irrigation consequences both pressing and politically daunting. The lack of a clear policy to integrate appropriate environmental and health safety measures in irrigation development planning is begging for a comprehensive solution. Mwangi examines the consequences of two major irrigation schemes on human health, as recent as February 1996 to February 1997 and concludes that the policy promises have not been translated into practice. Despite the accumulated knowledge about the consequences of fertilizers, pesticides and herbicides on human health, what became transparent here is a business scenario in which neither state nor business interests cared for the plight of those affected.

Many of the rural water supply development plans made by planners, government agencies and non-governmental organisations remain fruitless because their ideas cannot take root in new cultural landscapes. Social institutions, cultural norms, values, attitudes and practices are usually blamed for their failures, as there was no awareness of the hard-core resistance against changes of cultural self-image and identity. The result is that a significant number of water projects have either been abandoned or vandalised, or only function intermittently and have therefore failed to meet the expectations of the intended beneficiaries, governments and donors. Any meaningful and sustainable development of rural water supply must therefore address the inherent environmental socio-cultural set-up of the target area.

Based on the aforementioned observations, Mwangi examines the environmental and socio-cultural factors influencing the development, choice and use of rural water supply systems in Siaya District. To do this effectively, he has identified and critically examined the recipients' socio-cultural beliefs and practices, their water-use consumption and attitudinal variables (ability and willingness to pay for improved services) in relation to the development, choice and use of rural water supply systems. In addition, the nature and extent of community participation in the various stages of the project cycle were examined and the current rural water supply development institutions and technology used were assessed. The main criteria for assessment included feasibility, flexibility and acceptability within local environmental conditions. Mwangi's research findings have indicated that there was either a minimal or no community participation during the conceptualisation, planning, design, implementation, operation, administration, maintenance and management of water supply projects. Consequently, socio-cultural beliefs and practices, water-use practices and the local attitudinal variables were never integrated in the water planning process and the existing institutions and technology in use were found to be inappropriate since their design did not take complex environmental socio-cultural variables into consideration.

Despite decades of calls for popular participation in development, Wilson Mwangi's field experience and research findings show the contrary. Various rural water supply development strategies have essentially proved to be elitist and inconsistent with to the local communities' interests. Lopsided priorities, faulty technology choices and the adoption of inapplicable and irrelevant strategies have undermined the policy objectives and frustrated the water development efforts. Wilson Mwangi calls for the institutionalisation of an integrated alternative water-planning strategy taking into consideration indigenous values and traditional water resource management systems and based on meeting the people's needs, preferences, local initiatives and capacities. Sustainable rural water supply development can only be realized when peoples' participation is transformed from rhetoric to reality.

Nowhere in the environment-human interaction chain are the consequences of local environmental change on society more visible than in gender relations. The underlying assumption of Gariyo's chapter is that a woman role in capitalist development and the intensification of their sexual and social division of labour are historically determined. It is also the starting point in locating the forces that have contributed to the transformation of the conditions and character of women's production. Women's struggle to change their economic and social conditions, according to Gariyo, is part of a wider struggle to change the conditions of the rural economy and the social structure that supports it. In essence, the transformation of rural economies and the reformation cannot be accomplished without the social and historical forces responsible for the exploitation of women's labour.

Funna traces the differential consequences of the Structural Adjustment Policies (SAPs) of the energy crisis on the environment on gender with reference to the Kampala and Mbale Districts of Uganda. He found out that the structural reforms of the energy sector, put into force by the Ugandan government, have had varying effects amongst different rural agricultural producers as well as on urban households. While some households have gained, others have become vulnerable due to these policies and have subsequently lost. In conclusion, the energy crisis has had a more negative impact on the rural environment than on the urban environment, particularly women who headed households and who had no alternative to fuelwood, which is scarce and commercialised. The removal of subsidies for electricity coupled with the increase of domestic prices of crude oil and petroleum products reflected the fluctuations in the international energy market and aggravated the energy crisis in Uganda, according to Funna. This process has in turn contributed to a dramatic household shift from the utilisation of electric power and the rural nor the urban poor because electricity and kerosene could neither afford kerosene to charcoal and firewood. The shift to charcoal and fuelwood brought about increased exploitation of forest resources to meet the accelerating demand for charcoal and fuelwood. Subsequently, the increase in demand for charcoal and fuelwood as well as the increase in prices made the affordability of charcoal and fuelwood difficult for a majority of urban and rural dwellers. The energy crisis, made worse by falling incomes due to loss in demand for rural produce also caused retrenchment of large numbers of employees in the parastatal sector and the army.

The core question Maro's chapter attempts to answer is whether land management problems in Swaziland (land degradation, soil erosion, and overstocking) arise from inappropriate land use planning or usufruct and communal land tenure as the latter is sought to discourage investments in land. Or do problems arise from low producer prices and low rural incomes leading the peasants to concentrate on off-farm employment? To answer this question Maro has assessed soil and land suitability in three localities (Zombodze, Gege and Maphilenga) and found that farmers generally cultivate crops in soils they have intuitively identified as most suitable for agricultural production. There was no serious "mismatch" between land use: red and black loam soils ("sibovu") were correctly identified as suitable for maize and cotton, and stony soils and steep slopes were identified as unfit for cultivation but suitable for grazing. Therefore, farmers, through their accumulated experience, practise land evaluation and use planning, and by and large put land units to their suited uses. According to Maro, interventions through land use planning by government officials would probably have confirmed peasants' land use practices, but most importantly should integrate conservation and resource management (modern and indigenous) measures into a land use plans. Integrated land use planning is necessary not only to prevent further soil loss but also to institute measures that will maintain soil fertility under increasinglty intensive use. The dominant land management measures (animal manure, grass strips on contour bunds, some crop rotation, and rotational grazing) are inadequate and result in low crop productivity. Maro's chapter reveals that these land management and soil conservation measures are inadequate for sustainable production, hence the need to institute integrated land use planning for both arable and grazing areas is paramount.

Manyire's chapter provides an excellent linkage between a broader definition of environment, food security, and price and market liberalisation effects, from a gender sensitive perspective, on rural and urban Uganda. The main argument developed here is that in addition to structural differences in accessibility to food, the effects of price and market liberalisation may be gender differentiated. For although the market does not always operate against the interests of women, in the context of adaptation to rapidly evolving external economic conditions, the dual productive roles of women within national economies take on great importance. Available time after household provisioning and maintenance and differential access to economic opportunities and resources constitute the major constraints for the expansion of female accessibility to food. Furthermore, as substitute service providers and managers of the domestic economy, the burden of increased costs of food due to price and market liberalisation falls largely on women.

Aberra's chapter investigates pollution, as a major local environmental hazard not often associated with underdevelopment and low industrial output countries such as Ethiopia. Chapter tells a harrowing story reminiscent of the early industrial city. The poor residents of *Chile Sefer* neighbourhood suffocated with their clothes and bodies are blackened with soot. Soot emitted by the factory blackened not only the clothes that people wore, but also the laundry that they put out to dry. It is a case in which two of the main principles of environmental justice have been violated: 1) rights to protection harm and 2) the right of protection from harm and 3) the right redress inequality vis-à-vis geographic inequity and environmental marginalisation. Harrowing still is existence on paper, at least, of environmental policy instruments that beg for an urgent implementation. The implications of this chapter to environmental pollution control in Ethiopia and other developing countries are far-reaching. In fact it shows that the scale of industrialisation may contribute differentially to global change, but local communities take the full brunt of pollution and its accompanying environmental health hazards.

The United Nations Framework Convention on Climate Change (UNFCCC) Protocol was adopted at the third session in Kyoto, Japan, on 11 December 1997 and was opened for signature from 16 March 1998 to 15 March 1999. Less that two-thirds of the World states have signed the Protocol, although the Protocol kept silent about compensating local communities and more about global trading permits and emissions trade. The Failure of the Hague Conference (November 2000) on Climate Change and the lack of progress so far seems less worrying to those countries which possess the technological capacity and resources to protect their populations from some of the adverse consequences of climate change. The current debate on climate change and the furore surrounding preparations for the accession to the mechanisms embedded in the Kyoto Protocol and the United States and the European Union position in The Hague Conference is worrisome. This is particularly so in a matter of great importance not only for the industrialised world, but also for the developing countries. Mohamed Salih's chapter brings the grim realisation that beyond global environmental problem areas there are always victims who suffer the most. The chapter uses the case of household food security, which is contingent on the natural climate, particularly in countries where agricultural technology is so rudimentary to simulate favourable climatic conditions, to highlight the disastrous consequences on this particular sector. Dialectically, wetland reclamation is sought by developing countries households in order to compensate for the harvest lost during dry conditions. However, global Wetland Convention contravenes the interests of developing country households who have limited other opportunities to eke their living than reclaiming wetlands for crop production. Wetland reclamation on the other hand has contributed to significant biodiversity loss and the disappearance of valuable ecosystems and with them the also the edible plants and other species. At a larger synthesis, the consequences of climate change in the developing countries are more serious than we have ever thought and its consequences on other environmental problem areas such as food security; desertification, wetland disappearance and biodiversity loss, among others should not be underestimated.

The case studies provide a rare glimpse into the skewed relationship between the global-local nexus in three major global public policy domains and their consequences on local communities: 1) economic policy reforms, 2) environmental restructuring and 3) global public policy orientations in the fields of health and social exclusion. From the health impacts of irrigation to air pollution, climate change, gender, water and sanitation, land alienation and structural adjustment programmes (SAPs), local communities have bore the brunt of ill-conceived global projects. The case studies illustrate that the current debate on globalisation will be incomplete without balancing it with an equally passionate discussion about the consequences of globalisation on community and locality. Of particular relevance here, given the fact that these societies are dependent on the natural environment for their living, is the negative consequences of global environmental change for the local environments as well as the social, political and economic change they produce. These consequences are of at least two types: Those of a general nature and those of specific relevance to the variety of African conditions and local communities depicted by the case studies. This issue will be taken up in the conclusions below.

CONCLUSIONS

I argue elsewhere (1997) that the current debate on globalisation and the environment has evolved around four major themes, all informed by the multitude and complexity of a set of environment concerns dubbed even by most conservative commentators as symptoms of a global environmental crisis. From climate change to biodiversity loss and from air pollution to fresh water contamination, the symptoms of a global environmental crisis have often been treated from an essentialist position. This theme acknowledges the significance of the natural interdependence of the environmental life support system (air, water and soil) at a global scale. To that extent, the most common theme in the debate on globalisation and the environment is that, if it is environmental then it must imply global consequences on the human and the non-human world. Implicit in this assumption is that the impact of any large-scale development interventions (industrialisation, raw material extraction and processing, mega-irrigation projects, timber logging, mining, fossil energy exploration, production and processing and consumption etc.) in the environment would influence the quality of the environmental life support system. Such large-scale environmental interventions are part of an ecological modernisation process, which contributed to an unprecedented modification of the global environment. In essence ecological modernisation carries the insignia of globalisation and its capacity to produce momentous transformation in the way human lives are ordered and reordered, which in turn results in further modifications of the global environment. [1] Ecological modernisation can therefore be defined as manifestation of a complex and far-reaching development or process or modernisation in one of its most common interactions with the natural environment.

Some issues have emerged in this volume that show how local reality is interlocked into global processes and trends that enhance rather than reduce the production of environmental degradation and social deprivation. These issues are not specific to the case studies and are probably shared by a large number of the poor across the developing world:

Firstly, the case studies warn against the current obsession with global environmental change and secondly against the neglect of the local environmental imperative. In the course of collecting material in preparation for the writing of this introduction, I became more convinced that social and natural science polarisation still haunts the ability of both to be able to advance a genuine cooperation based on each others comparative advantage. Studies on soil erosion, forestry, pollution and wetland reclamation are still the monopoly of natural scientists, with a token input by social scientists in order to break through the interdisciplinary conditionality imposed by some research funding institutions. The main reason is that the scientific verification of the extent of any problem is often locality-specific (for more on this see Idris 1997).

Social scientists have, on the other hand, strived to match natural scientists' obsession with global environmental change and most recent writings are dominated by concerns of questions pertaining to environment and globalisation. Although global trends are legitimate areas of inquiry, the obsession with global environmental change has not been matched with equal interest in local environmental change and its consequences on society. This has subsequently; 1) drawn attention away from community and locality to a notion of global environmental change often removed from the reality of those at the receiving end of these changes. 2) Larger (global) synthesis obscures local reality, and imposes conditions often insensitive to local differences, needs and production systems, even distorting them. The case studies reveal the seriousness of the difficulties involved in conflating national policies and local reality, let alone global policy instruments on local communities. 3) The development of technologies capable of detecting global environmental change has not been matched by similar efforts to develop affordable technologies appropriate for investigating local environmental change and is still a long way further up the road. 4) Global environmental governance and institutions are still dominated by global economic powers that have found it increasingly difficult to deal with the multitude and varieties of local realities unleashed by global environmental change. Hence problem avoidance and the relegation to oblivion of the difficult parts (local) of the global equation have become common practice. Ironically, the global void has become more accessible to some researchers than the majority of impoverished humanity that populates the planet earth.

---

[1] Optimists such as Hannigan (1995:184) argue that, ecological modernisation and technological advancement will contribute to the annexation of time and space at a global scale through the dematerialization of production (virtual work-place etc), reduce transport and allow people the opportunity to work at home.

These concerns have also been reflected in an earlier work in which research findings on the performance of environmental planning and policies in nine African countries (Botswana, Ethiopia, Kenya, Lesotho, Malawi, Sudan, Tanzania and Uganda) was presented (Mohamed Salih and Shibru 1999). I have shown in the introduction to that volume that (Mohamed Salih 1999b: 10-11) concerns with adapting national environmental action plans and national conservation strategies to global conventions and from there to local communities have not produced the desired result. I concluded that:

1. Since the late 1970s, regional and national environmental policies in the African region are increasingly influenced by global environmental forums, negotiations, conventions, treaties, declarations and the emergence of the concept of global environmental governance (Commission on Global Governance 1995).
2. The existence of governmental and non-governmental (NGOs) institutions involved in the environmental policy process, in implementation, and in monitoring. Most countries involve such institutions in public debate on key environmental issues and concerns. These institutions are weak and operate on a sectoral basis, with inadequate co-ordination and overlapping responsibilities. They also suffer from inadequate knowledge and limited information about the extent of resource degradation as well as the sound implementation of resource management policies.
3. The integration of environmental impact assessment (EIA) as an anticipatory-preventative environmental policy instrument. This relates particularly to the insistence of international finance institutions (such as the World Bank) and bilateral development agencies on better cost-effectiveness and distribution.
4. Owing to underdevelopment and market weakness (inadequate pricing, lack of trained human resources and institutional constraints), fiscal and socio-economic incentives are very rarely used as environmental policy instruments. This is an area where human resource development is desperately needed if African environmental policies are to be translated into powerful resource management tools.
5. Instead of economic instruments, legal instruments are used, some of which date back to the colonial period. However, regional and national environmental laws are increasingly influenced by international environmental laws proposed and drafted by global environmental forums, negotiations, conventions, treaties, agreements and declarations, and the emergence of the concept of global environmental governance (Commission on Global Governance 1995).
6. Legal environmental policy instruments promulgated within a dysfunctional regulatory framework are insufficient and suffer from inadequately defined objectives and lack of trained personnel.

Understandably, local communities cannot put their faith in environmental policies and plans imposed on them by global or national institutions that often deprive them of access and control over their local environment.

The specific policy implications of the case studies can be summarised in at least four points: First, in rural Africa state authority is actually not as pervasive as has often been portrayed. Moreover, local communities maintain a certain degree of autonomy, norms, values and beliefs that have provided refuge from alien and often incompatible resource management values imposed by the state. Local values and norms have become markers of political identity resistant to state hegemony and unwanted interventions. A situation has emerged where state policies are forced to co-exist alongside local institutional autonomy. The inherent duality of two, apparently different perceptions of natural resources is mediated by a deliberate configuration of peoples' livelihood, socio-economic and environmental conditions in order to come to terms with two distinct forms of production. The production of cash crops in the state dominated economy and the production of food for subsistence and local reproduction. In the circumstances, local compliance with local norms is more important and pervasive than compliance with state policies and alien development norms and values. The gulf between the two tells a story of a mounting tension between state illegitimacy and local self-assertion. However, the institutional arrangements between the state and local communities have come to a direct confrontation when the state intervenes directly in the appropriation of local resources on behalf of public or private interests. In this latter case, duality is transformed into hostility and conflicts, and as is often the case without local acceptance, national environmental policies are doomed to failure.

Secondly, in urban Africa, the environmental and livelihood conditions are reminiscent of the housing, food and sanitation problems experienced by the early industrial cities in Europe, as depicted in Goldblatt (1994: 60) where, high density and stationary living gave way to:

> "The relentless accumulation of human excreta and the by-products of consumption. Not only do these constitute a physical obstruction but, if not properly dealt with, pollute what remains of the ecosystem in a city and/or create new micro-habitats for a range of pathogenic micro-organisms, insects and mammalian vermin. It is hardly surprising, therefore, that the environment of the industrial city was profoundly unhealthy, with its polluted water supplies and streets, cesspools, saturated land, garbage dumps and dirty human bodies".

This is a vivid reminder of the situation in many urban cities in the developing world today. There is much to be learnt from how the early industrial city was able to cope with some of these environmental problems. Not only questions of costs and benefits can be posed; environmental policy historians could rightly pose questions pertaining to the socio-cultural impediments that might have constrained urban production and health policies. While reinventing the wheel is neither creative nor constructive, assuming that the current socio-economic conditions prevailing in most of Africa could support a similar level to today's post-modern industrial city is pretentious and counter-productive.

Thirdly, the concern with good governance in politics has also not been matched with an equally pressing need for good local governance and institutions capable of implementing some of the local implications of global and national environmental policies. The weakness of local environmental policy institutions that have been distorted by centuries of colonization and decades of authoritarian post-independent states is all too apparent. In the African conference on Rio+5 held in the Egyptian capital Cairo, the delegates of 16 African countries adopted a resolution, which precisely emphasised that

Local communities' institutions and resource management systems are vehicles for internalising empowerment, participation and accountability ethos in society. On the contrary, national level planners and politicians serve their political whim rather than empowering the local communities, which are often uses as political resources in their venomous struggle for power. In fact, participation, empowerment, gender sensitive development and environmental management have increasingly been normalised and used as instruments to co-opt the very communities that are supposed to be empowered to participate in development and environmental management.

Fourthly, the ultimate goal of empowering local institutions is to revive local self-reliance in the face of the economic crisis that has beset the African continent (Gooneratne and Mbilinyi 1992) as well as the care for the community. Unfortunately, some very basic questions over which people have laboured for several development decades including gender differentiated access to land, genuine local participation and respect for local initiatives are yet to find satisfactory answers. Our obsession with the global, as I have tried to show in this chapter has to some extent undermined our ability to engage the local and how such some such global developments might have in reality impaired the capacity of local communities to survive in a production environment.

In short, sustainable local environmental management is a prerequisite for reducing poverty, gauging people's trust in their capacity and the capacity of their environment to deliver sustainable livelihood conditions. One can hardly escape the fact that global environmental change is a symptom of much deeper social, economic and political problems expressed in local environmental change. The dialectical relationship between the global-local nexus is far more complex than we often tend to believe.

REFERENCES

Adams, W. (1992) *Green Development: Environment and Sustainability in the Third World*. London and New York: Routledge.

Africa Development Bank (1998) *Africa Development Report*. Oxford: World Bank and Oxford University Press.

Barbier, E. B. (1991) "Natural Resource Degradation: Economics and Management", in James, T. Wimpenny (ed.) *Development Research: The Environmental Challenge*. London: Overseas Development Institute, pp. 43-57.

Basu, A. and Steward, F. (1995) Structural Adjustment Policies and the Poor in Africa: Analysis of the 1980s, in F. Stewart (ed.), *Adjustment and Poverty: Options and Choices*. London and New York: Routledge, pp. 138-170.

Cheru, E. (1992) *The Silent Revolution in Africa*. London and New Jersey: Zed Press Ltd.

Chambers, R. (1988) "Sustainable Rural Livelihoods: A Key Strategy for People, Environment and Development", in C. Cornoy and M. Litvinoff (eds.) (1988) *The Greening of Aid: Sustainable Livelihood in Practice*. London: Earthscan Publications, pp. 1-17.

Davidson, J. (1991) "Gender and Environment: Ideas for Action and Research", in James, T. Wimpenny (ed.) *Development Research: The Environmental Challenge. London: Overseas Development Institute*. Pp. 137-144.

Goldbblatt, D. (1996) *Social Theory and the Environment*. Cambridge: Polity Press.

Goonerrante, W. and Mbilinyi, M. (eds.) (1992) *Reviving Local Self-Reliance: People's Responses to the Economic Crisis in Eastern and Southern Africa*. Nagoya: United Nations Centre for Regional Development.

Kikula, Idris S. (1997) *Policy Implications on Environment: The Case Study of Villagisation in Tanzania*. Uppsala: Scandinavian Institute of African Studies.

Kousis, Maria (1998) “Ecological Marginalisation in Rural Areas: Actors, Impacts, and Responses”. *Sociologia Ruralis*; 38, 1, April, pp. 86-108.
Martell, L. (1994) Ecology and Society: An Introduction. Cambridge: Polity Press.
Meadows, D. H., Meadows, D. L. and Randers, J. (1992) *Beyond the Limits: Global Collapse or A Sustainable Future*. London: Earthscan Publications.
Mohamed Salih, M.A. (1999a) *Environmental Politics and Liberation in Contemporary Africa*. Dordrecht, Boston and London: Kluwer Academic Publishers.
Mohamed Salih, M. A. (1999b) “Introduction: Environmental Planning, Policies and Politics in Eastern and Southern Africa”, in M. A. Mohamed Salih and S. Tedla (eds.) *Environmental Planning, Policies and Politics in Eastern and Southern Africa*. Basingstoke and New York: Macmillan and St. Martins, pp. 1-17.
Mohamed Salih, M.A. and Tedla, Shibru (eds.) (1999c) *Environmental Planning, Policies and Politics in Eastern and Southern Africa*. Basingstoke and New York: Macmillan and St. Martins.
Mohamed Salih, M. A. (1997) "Global Ecologism and Its Critics", in Thomas, C. Peter Wilkin (eds.) Globalisation and the South, St. Martins, New York, Macmillan Press, Basingstoke.
Mohamed Salih, M.A. (1994) ‘Introduction’, in M.A. Mohamed Salih (ed.), *Inducing Food Insecurity: Perspectives on Food Policies in Eastern and Southern Africa*. Uppsala: Scandinavian Institute of African Studies.
Mosely, P. (1994) ‘Decomposing the Effects of Structural Adjustment: The Case of Sub-Saharan Africa’, in R. Van den Hoven and F. V. Kraaij (eds), *Structural Adjustment and Beyond in Sub-Saharan Africa*. London: James Currey.
Shepherd, Gill (1991) "Tropical Dry Forests", in James, T. Wimpenny (ed.) *Development Research: The Environmental Challenge*. London: Overseas Development Institute, pp. 61-65.
Stewart, F. (1995) *Adjustment and Poverty: Options and Choices*. London and New York: Routledge.
Tallegen and Wolsink (1998) *Society and its Environment: An Introduction*. Amsterdam: Gordon and Beach Science Publishers
UNICEF (1991) *Challenges for Children and Women in the1990s*. Nairobi, Kenya
World Bank (1994) *Towards Sustainable Development in Africa*. Washington DC.

2 Human Health and Irrigation Development in Kenya

Samson Wokabi Mwangi

The human health threats posed by irrigation development have made it necessary to understand the consequences of local environmental changes on the economic and social well-being of people. Local environmental changes and the human health risks associated with them are commonly attributed to the improper understanding of the complex processes involved in the construction of irrigation systems. In this respect, while it is necessary to recognize that the environmental and health problems associated with irrigated agriculture constitute an unnecessary evil, such schemes have often been constructed not without good intentions. Such intentions include the interest of societal welfare at large, bearing in mind that the negative consequences could either be avoided or minimised by introducing appropriate corrective measures.

Therefore, it has become necessary to analyse the dynamics of irrigation development and understand their implications in a given local environment. It is important to delineate the frontiers of irrigation development potentials as well as the diagnosed environmental and health problems. Corrective measures should be incorporated in the design, execution, operation and maintenance and long-term sustainability should be aimed for.

Attention should therefore be paid during the planning stage of a project to ensure that environmental and health safeguards are properly taken into consideration. Engineers, planners and managers more than any other group of participants are involved in the development of irrigation schemes. These experts, however, are not specialists in medical, social, economic, environmental and health impacts associated with irrigated areas. Issues and problems related to medical and/or environmental health in irrigation schemes are outside the scope of their responsibilities. For them, information on the impact of irrigation projects on medical and environmental health for the beneficiary population is usually difficult to document. As a result, the negative environmental impact of irrigated agriculture remains undocumented or poorly recorded. However, it is only recently that development professionals have begun to recognise the importance of environmental factors. Also most development agencies that fund and/or are involved in implementing irrigation projects are introducing procedures, which they believe will ensure that serious problems are avoided.

The policy of the Kenyan government on irrigation emphasises that land reclamation by drainage and flood protection are the most realistic irrigation measures for increasing agricultural productivity (GOK, 1994). The other aim is to

M.A.M. Salih (ed.), Local Environmental Change and Society in Africa, 2nd Edition, 17–40.

expand cultivable land to meet the ever-growing demand for food by restructuring the country's agricultural sector, settling people in 'new lands' and creating jobs. This policy orientation reflects the fact that about three quarters (about 75%) of the total land in the country is arid and semi-arid. The seventh national development plan stipulates that the government policy on irrigation will favour commercially oriented smallholder group-based irrigation projects able to meet their capital investments and operation costs (GOK, 1994:141). Adverse environmental and health problems identified by the current plan exacerbate low-level productivity because they are not analysed adequately before implementation of irrigation projects.

It has been made a requirement that environmental and health impact assessment be undertaken before the implementation of irrigation projects. Yet, there are no criteria specifically developed for use in the assessment of Kenya's irrigation and implementation planning. Out of 44.6 million ha of irrigable land, 36 million ha is yet to be put under irrigation. Not until enough information has been provided about the relationship between expanding irrigation activities and the type and magnitude of environmental hazards and health risks involved, may Kenya proceed with confidence to expand her agriculture production programmes via irrigation.

This chapter explains the magnitude of local environmental change and the resultant health problems associated with the development of irrigation schemes in Kenya. It reveals that despite a world-wide accumulated knowledge of the diverse effects of irrigation on human health, Kenya has not undertaken serious policy interventions to incorporate corrective measures in designing, implementing and managing such projects in order to reverse the adverse environmental and health consequences of irrigation.

PROBLEM ANALYSIS

Although several researchers acknowledge the likely impacts of irrigation development in arid and semi-arid lands, much of the scientific knowledge gained is discipline bound, disregarding the complex interactions within irrigation ecosystems. Retrieval of diffused available literature on environmental implications of irrigation development has always been a formidable task. Attempts are here made to review available literature.

Irrigation is generally defined as the application of water to soil to provide a favourable environment for the growth and development of crop plants (Aral, 1971). Researchers in different parts of the world and especially social scientists have studied the impact of irrigation development. These studies are of a different nature and of different levels i.e. at the household level or at the macro level. The ultimate aim is to bring out the impact of irrigation development on the various factors or aspects of development either at the command area or the regional area, either at the national or the international levels. The present study focuses on the environmental and health problems and planning perspectives in an irrigation settlement with a view to coming up with an environmental planning and management methodology, which could be used to minimise the negative consequences identified.

Worthington (1976) notes that the full impact of irrigation and the associated drainage development upon water, soils, aquatic and terrestrial ecosystems and human health extend through the social and economic fabric of local and national societies. According to White (1978), well-planned and administered irrigation projects can be of enormous benefit to developing countries, not only by increasing food supply, but also by contributing to the overall economic development. A successful irrigation scheme, he contends, can improve the well being of local populations by providing an abundant and widely distributed water supply. Better nutrition and planning of housing, health care facilities and a general increase in the standards of living can also be seen as impact possibilities of irrigation development. Irrigation activities in MIS have led to an apparent increase in agricultural productivity.

According to Rajappa (1981) cited by Biswas (1984), irrigation schemes often solve drinking water problems at least when piping via tube wells is not possible. Any such improvement is of major public health importance, and, provided some attention is paid to the question of potable water, excellent results are obtained at minimal expense. On the other hand, where basic precautions are neglected, a great opportunity is lost. If properly carried out, provisions can also be made for waste disposal facilities. Hence, Rajappa concludes that, engineering works done preparatory to the setting up of irrigation schemes can be beneficial to the local population and any future inhabitants by making the area less prone to environmental and health hazards.

Chen *et al.* (1980) cited in Arnon (1981), observes that where irrigation authorities deal with the construction of new villages, there is an opportunity to design new houses in accordance with minimal sanitary standards, as well as with the preferences of the prospective inhabitants. This can help keep the resettlement stress within tolerable limits. The presence of adequate quantities of water is one of the prerequisites for satisfactory personal and domestic hygiene, and it plays a major role as a background for the control of communicable diseases. In many arid localities, a sharp drop in incidence of trachoma and scabies has been noted after establishment of an adequate water supply through an irrigation scheme.

Irrigation can potentially greatly increase agricultural output; whether or not the individual cultivator benefits from any such increase depends very much on the style of irrigation management. Rao (1987) argues along the same line, that irrigation development plays a major role in rural development by transforming a once dry village to prosperity. The author undertook a comparative socio-economic study of two areas where in one area there were irrigation activities while in the other these activities were absent. Both areas had the same bio-geophysical characteristics. He was of the opinion that irrigation leads to the improvement of agricultural productivity, which eventually leads to rural development. His analysis, although dealing fully with the economic benefits of irrigation to rural development, compares well with the present study where a comparison has been made between an area with irrigation and another one without to reduce the chance element.

Attempts have been made to ascertain the impacts of irrigation development but little has been done to review the existing planning and management strategies in these areas and to introduce appropriate changes, which would help increase the advantages from irrigation development.

Even with the best planning, some important factors could be overlooked and unexpected environmental impact and socio-economic impact will sometimes be generated. Careful review of the past experiences should reduce the risk of this happening and should give planners a chance to prepare contingency plans for any problems which may arise. For instance, according to Kovda (1980), irrigated fields are very complicated anthropogenic ecological systems that include the following interrelated components: land, climate, water, soils, plants, animals and micro organisms, machinery, fertilisers, biocides, human activities and programmes, crop yields and wastes. The irrigation agro-ecological systems are therefore much more anthropogenic and complicated than the agro-ecosystems of dry agriculture even more so of pasture ecosystems.

Unforeseen negative after-effects may tell on all the elements of the environment both in the form of rapid changes occurring and unforeseen variations in the hydro-ecology of land manipulation of soil and landscape ecology. Chancellor-Weale (1991) contends that irrigation projects no longer enjoy the automatic approval generally accorded to them earlier in this century. Poor returns, high initial and recurrent costs, and development impact much lower than had been forecast have reduced the popularity of irrigation projects. The author argues that, in the worst cases, considerable environmental damage has occurred and that the expected beneficiaries have abandoned some schemes.

White (1978) notes that in most cases, and especially in large plains, the water table is raised by irrigation, and capillary arising from the higher ground water increases the accumulation of salts both in the ground and in the near surface water. According to Schilifgaarde *et al.* (1974) cited in Hotes and Pearson (1977), the total rise in salts due to irrigation normally accompanies rising ground water. Soil deterioration occurs widely due to the very low efficiency of the majority of irrigation systems, which may average up to 30-50% (White 1978).

The danger of water logging of irrigated lands persists even when improved technology and better equipment are applied in the irrigation system. Water logging is a common feature of the majority of irrigation systems around the world (Kovda, 1980). The rise in the ground water may in some cases be rapid, reaching several meters a year, but usually it is measured in centimetres a year. This process is inevitable whenever drainage is inadequate.

Closely related to the problem of water logging is the salinity problem. Salinity, according to Kovda (1980) is the concentration of salts dissolved in water or soil, perhaps due to the dominant presence of a particular salt or several salts. Groundwater always contains some salts and its saltiness is in proportion to the soluble salts present in the soil and sub-soil. Most cases of secondary salinity are therefore associated with the rise of the water table.

Hotes and Pearson (1977) have observed that adverse effects of irrigation on water quality have been identified as increases in salinity, turbidity, colour, taste, temperature, nutrients, bacteria and viruses all causing economic losses and deterioration to the environment.

Part of the irrigation water goes back into river networks or groundwater systems and with the return flow it may once again be used for irrigation or as domestic water (Sharkar and Kurma, 1993). Pollution caused by water bodies from the return flow is as important as it poses real dangers to aquatic inhabitants, human or animal.

Experience has shown that the establishment of an irrigation scheme may render a region more vulnerable to diseases (WHO, 1980; FAO, 1987). According to FAO (1987), water can carry toxic chemicals as well as many communicable diseases, serving both as a transfer medium and as a habitat for vectors and intermediate hosts. The major diseases are transmitted by mosquitos and simulium, and borne by snails. Diseases not existing in an area before may appear after irrigation activities have been introduced. Low prevalence and minor intensity can grow to persistent and massive infectiousness. Severe and debilitating infections may then reduce the workers capacity with socio-economic consequences. Malaria, schistosomiasis, amoebiasis, dysentery and other vector- and water-borne diseases have caused the greatest damage in countries where irrigation has been developed with no consideration for the effects on public health (WHO, 1980). Singh and Afroz (1985) noted that the main purpose of irrigation development is to improve the well-being of the community. If well-planned research policies are not drawn up at the initial stages of irrigation development, this purpose will be defeated. In the absence of a timely public policy, irrigation development may lead to unintended consequences.

In concluding the foregoing discussion, it is obvious that irrigation brings about a series of changes in different components of a command area. Such changes can be categorised as stable or unstable and furthermore as controllable or uncontrollable. Stable changes are sure to occur (many studies have focused on this) but unstable changes may or may not occur, depending upon the combination of other natural factors. Controllable changes have a definite magnitude and tendency that can be determined by society. Uncontrollable changes cannot be so determined (Sharkar and Kurma, 1993). As a rule, stable uncontrollable changes have a beneficial impact on the environment (Johl, 1980).

Therefore there is a need for careful and unbiased appraisal in order to identify which approaches to irrigation development and which management strategies offer the greatest benefits at the lowest cost. More care is needed to assess in advance the environmental and socio-economic effects of irrigation development. Assessments, thorough or otherwise, are either made too infrequently or just for the sake of making them. The present study attempts to ascertain the environmental and health problems of irrigation development in two NIB schemes. Existing planning and management strategies will be reviewed, and finally appropriate planning and management strategies will be given for the existing irrigation scheme as well as for future schemes. From the above reviewed studies, environmental and health problems of irrïgated agriculture have been seen. It is necessary after identifying these problems to give appropriate planning and management recommendations to mitigate, control and contain the problems identified. The comparative analysis will help in eliminating the chance element.#

IRRIGATION DEVELOPMENT PLANNING

The long history of irrigation development has not always recorded success. Some past schemes and some very recent ones have suffered from severe environmental deterioration (in their environments) (White, 1978). A number of irrigation projects have also been affected by socio-political changes. Some have proved excessively expensive in relation to their economic returns. Still in others, some of the farmers have abandoned their land or have never felt at home in their new setting. Any sober assessment of the opportunities to reap benefits from irrigation must take stock of the reasons for past failures as well as the resources available for future development (White, 1978).

The development of new irrigation schemes and the expansion of old ones is a positive and commendable step toward the overall elevation or improvement of nutritional, social and economic status. As many writers have noted, health, social and environmental problems accompany such development. Planning of major as well as minor irrigation schemes has tended to focus on achieving the most technically efficient system of water distribution and taking less account of managerial, social and environmental factors. Inadequate socio-economic and environmental investigations have caused the irrigation planners to fail in allowing for adjustments in irrigation induced unforeseen consequences.

According to Carward (1991), planning for irrigation development is already a lengthy and complex process covering an array of social, economic, technical and engineering factors. Given this complexity, it can be said that, one should always hesitate including issues like environmental and health implications in feasibility studies.

Currunthers (1978) admits that in general irrigation schemes have been a terrible failure and observes that, the scant information on performance suggests that the gap between potential and realised benefits has, in fact, increased. The main reason for this, he asserts, is that participants do not integrate their professional activities according to an agreed conceptual understanding of the problem. Currunthers further discusses the pressures applied to the planning and operation of an irrigation scheme and lists the groups that apply such pressure. Each group, the states, has its own preoccupation. The agronomists are concerned with crop yield or the returns of the land from modern irrigation development. Engineers will seek to satisfy their vision through new dimensional structures whereas planners and administrators look at the gains from managerial control point of view. Aid donors and politician usually have a very short-lived view. Currunthers continues, "it always strikes me as ironic that we have placed the responsibility for curbing our 'defective faculty' for the future in the hands of individuals whose prime interests rarely exceed five years".

Politicians, economists and engineers have generally dominated irrigation development planning and management and success has commonly been judged according to economic and technological criteria. Predictions of the impact on the environment and health may even be seen as the work of 'soft science" and therefore get minimal attention. The lessons of operating experiences are largely ignored in irrigation development planning. The author concludes by saying that studies must be speeded up and funds made available to pay for them and wherever possible indigenous planners, rather than foreigners should be used to do the job. The planning process therefore should involve the people at the grassroots level who will be affected by irrigation development projects. Planners' targets are rarely met: the overall productivity of water is much lower than might be expected. The surface-water delivery system and the patterns of distribution are often extremely inequitable, with farmers in the head-reaches receiving more than those at the tail. Planning and design processes tend to be dominated by technical experts especially engineers, with economists called to calculate the costs and benefits of what is technically but not socially or environm*entally* feasible.

Thitai (1990) notes that the ultimate success of an irrigation scheme depends on an integrated approach which considers all aspects such as increased agricultural output, human resources, public health and environmental quality. However, in most cases it has been noted that the introduction of environmental and public health element at the planning stage involves only a negligible increase in the overall costs of the project. These, it is always assumed, could be undertaken later in the absence of proper provisions. In most cases, if not all, this element is only remembered when the damage done to the quality of the environment has reached a threshold.

The successful development and utilisation of an irrigation scheme are so intimately related to community interests and activities that they should be planned on a broad basis. Considerations should be given to the available land and water resources, climate, crops, people, health aspects and the effects of other water use developments (WHO, 1965). The planning of an irrigation scheme should take into account the prevalence of human and animal diseases, vital statistics and available scientific data on environmental problems applicable to irrigation schemes, organisation, engineering services, operation, management and economic and marketing factors. It should be noted that the planning organisation finds this a great array of activities which are not of immediate importance and are therefore generally forgotten. For instance, according to WHO/FAO/UNEP (1990), the costs of molluscicides should be considered in the planning and design of irrigation systems, especially since corrective measures can be used to reduce the costs per acre.

According to Chancellor-Weale (1991), failures and problems in irrigation settlements stem from the fact that initial planning and project selection gives little priority to the sustainability of irrigation development. The reliance on Internal Rate of Return (IRR) as a single figure assessment by which alternative irrigation development proposals can be compared, has encouraged the selection of proposals with a deleterious effect on the environment.

Two related problems therefore complicate the process of irrigation development planning:

1. Large number of interest groups each with varying affinities and approaches to the subject. These groups promote irrigation because it is consistent with their view of the needs and priorities of development.
2. All irrigation development projects have had to satisfy a large number of objectives some of which are consistent while others conflict. Assessing the trade-off between the various levels of achievement of multiple objectives is one of the activities in irrigation development planning (Currunthers and Collins, 1981).

The health aspects of irrigation development have typically been neglected, and impose social costs upon communities dependent on irrigation. Irrigation development often means rapid building of townships to house construction workers and sometimes displaced families. Preventive public health measures such as sanitary facilities, water supply and immunisation can help at this stage. In the operation phase there is more surface and perennial water which, with a dense population, may lead to unhealthy conditions. Adequate drainage is important for control of insect and snail populations, which affect the incidence of several severe diseases. It is therefore very important that a multidisciplinary team be involved in the planning and design of new or the rehabilitating of old irrigation projects.

The above studies show the complexity of irrigation development planning and will be relevant to the present study which attempts to give an environmental planning and management methodology. It is hoped that the reviewed studies have identified interest groups in irrigation development, and subsequently broaden the scope of elucidating the factors associated with the health consequences of irrigation schemes in the particular case of Kenya.

IRRIGATION DEVELOPMENT AND ENVIRONMENTAL HEALTH

In 1981, the Ministry of Agriculture established a policy with regard to an accelerated programme on irrigation and drainage development in Kenya. It emphasises that irrigation, land reclamation by drainage and flood protection are major and attractive alternatives for increased agricultural productivity. Other aims are to expand cultivable land to meet the ever-growing demand for food and to restructure the country's agricultural sector as well as to settle people on new lands and to create substantial employment opportunities.

Presently some 54,000 ha are estimated to be under irrigation, representing only 10% of potential irrigable land (GOK, 1994). Compared to the existing high- and medium-potential land estimated at 8.4 million ha, the irrigation potential of 0.54 million ha is rather limited and the existing irrigated area of 54,000 ha is totally insignificant. There is therefore an urgent need to expand irrigation activities in these potential areas. The government will in the long run strive to develop the irrigation potential of the country through improvement of planning and implementation processes as well as through their operation and maintenance.

The Seventh National Development Plan (1994-1996) stipulates that the government policy on irrigation will favour commercially oriented smallholder group-based irrigation projects able to meet their capital investments and operation costs. Useful experiences gained in construction, operation and maintenance of any major infrastructure will be used in the development of new schemes.

Adverse environmental and health related issues have been identified as exacerbating low-level productivity because they had not been analysed adequately before implementation of irrigation projects. Efforts will therefore be undertaken to set prerequisite mandatory minimum environmental and health impact assessment criteria before implementation of irrigation projects (GOK 1994). The foregoing review shows that the government has realised the adverse effects of irrigation development and the need to mitigate against them. However, it has been noted that there is a lack of adequate data for/of such a policy. It is assumed that the present study will assist in the implementation of the government's policy in solving irrigation induced environmental problems.

TOWARDS A CONCEPTUAL FRAMEWORK

The conceptual framework that guided the research is borrowed from the principles of sustainable development. This concept tends to view both economic and social elements on the one hand and ecological aspects on the other as interrelated. All these elements and aspects are in a continuous process evolving into a complex system. Irrigation ecosystems tend to be complex and their planning requires the integration of all these components.

Irrigation development involves an interface between the natural environment, technology and human aspects and may have some dysfunctional consequences on the intended beneficiaries if appropriate measures are not taken at all in the planning stages. The uses of these biophysical resources need to incorporate measures safeguarding them from the degrading consequences of human activities. In order to do this, the integration of ecological principles and concepts in the use of resources is central to the sustainable development paradigm. The principles of sustainable development require that future generations must be able to use the available soil and water resources to feed and support themselves. Preventive public health measure such as adequate drainage and sanitary measures must be put in place while agricultural productivity should at the same time be increased.

The interactions and linkages between different elements and aspects in an irrigation ecosystem show that all these parts should function in harmony to maintain the whole. The other theoretical aspect used in relation to sustainable development is borrowed from the General Systems Theory (GST). The use of GST applies to environmental planning. In the context of GST, an irrigation system as made up of different parts that function in harmony to maintain the whole system. At the same time, the GST explains the working of ecological systems at the national and regional levels, and even the global level - the biosphere. Any disruption in any part of the system will eventually lead to disruption of the operation of the system. This theory when utilised in the planning and management of an irrigation settlement takes into consideration the interrelatedness of the different components that irrigation systems in arid lands are made up of. This could only be done when a multidisciplinary team of economists, agronomists, environmentalists, engineers, politicians, sociologists and public health officers are involved in the irrigation development planning process. The approach to the environmental issues in the two irrigation schemes has been developed from a stand point that considers a whole range of households in the context of both environmental and health problems and the existing planning and management to deal with these problems. The intersection of these problems as well as the existing planning and management issues help to improve our understanding of the theoretical and practical issues that fall within the scope of today's environmental issues related to irrigation development. Environment is studied by considering the following variables: Demographic and family characteristics, housing conditions, water supply and sanitation, energy sources and uses, human health, environmental problems as perceived by farmers, and soil and water characteristics. Enquiries into the existing planning and management arrangements have supplemented data collected on this complete set of variables.

IRRIGATION AND ENVIRONMENTAL HEALTH IN PRACTICE

In this chapter, findings based on the interviews of 324 respondents, 55 respondents of the 2 NIB irrigation settlements as well as interviews with key informants are presented. Data synthesised from field observations and secondary sources have also been included.

The information presented in this section stems from various sources. Specific data inventories in the present study comprise three categories:

(a) Data collected through the scheduled interviews of a sample of 324 individuals and 55 household heads in Mwea and West Kano respectively;

(b) Data collected through dialogue with scheme management staff and other selected key informants and
(c) Data collected through the researcher's field observations and secondary information from existing data.

The data collected has centred mainly on the various aspects of the physical and human environments. Emphasis was placed on the explanation of data within the context of environmental and human problems in the two schemes. Preference in the selection of data has been given to purely quantitative variables and data based on qualitative scales has also been considered. Environmental and health problems revealed by these data are explored and explained in detail.

To collect accurate data on these problems, a household survey was conducted by use of an interview schedule. A macro survey was also conducted using field observations and secondary sources of data. Environmental and health problems identified are presented at these two levels.

The size of families was found to be fairly high, the majority counting more than 6 persons per family in the two NIB schemes studied. It was found that 60% and 58% had more than 8 persons per family in Mwea and West Kano schemes respectively. The problem associated with this situation is that the original number of people settled has increased more than fourfold. There is therefore an increasing population pressure which calls for an urgent need to expand the settlements. Congestion in the villages also causes the transmission of communicable diseases like cholera.

Respondents interviewed were allocated a 50ft by 80ft plot of land on higher grounds for the purpose of building houses. The results of the household survey showed that the majority of the respondents live in iron-roofed houses. Almost all types of houses used by tenants are characterised by factors which contribute to poor health. The survey showed the presence in the human environment of pathogens because of a lack of basic infrastructure and services like sewers, drains or services to collect solid and liquid wastes and safely dispose of them. There is a lack of safe and sufficient water supply. Overcrowding and cramped living conditions, as observed in all the villages, increases the risk of transmission of airborne infections and increases the risks of accidents. Many diseases such as tuberculosis, influenza and meningitis are therefore easily transmitted from one person to the other.

Housing conditions are determined by the nature of the house, the number of rooms and the household size. It was observed that the exclusive concern of the NIB, the board that manages these irrigation settlements, was for a high level of technical achievement which has led to the neglect of the problems of human welfare. As a result, houses are overcrowded, smoke-filled and lack privacy. The physical expansion of the two schemes seems to have reached a limit and the management should seriously look into the welfare standards of the tenants. Apart from the social and health implications these conditions can be economically harmful, discouraging full participation in the scheme and commitment to long-term investments which in turn may lead to poor performance of the irrigation schemes.

The results of the interview survey revealed that the greater majority of the respondents obtained water from nearby canals. This water is of poor quality and has been contaminated with pathogens. The poor quality of water can be attributed to subsurface return flows from the irrigated fields. The majority of the tenants interviewed defecated at the water points within the schemes. Considering the residents' bathing and domestic use of water, they have become extremely vulnerable to water-borne diseases. It has been observed that in these villages some NGOs had assisted in constructing boreholes.

Within the schemes, improper human waste disposal can also be held responsible for water-borne diseases. All the respondents interviewed had pit latrines as the only method for excreta disposal and a large percentage had shallow pit latrines ranging from between 2 to 6 metres deep. The majority of the pit latrines were only 5-10 metres away from the houses because of the limited size of the plots. The absence of drains and sewers to take away wastewater and rainwater from the settlements leads to stagnated pools harbouring mosquitoes. The lack of sewerage connections and garbage collections may lead to critical environmental problems. A lack of readily available, treated drinking water, of sewage connections or other systems for the hygienic disposal of human waste, of garbage collection and any basic measures to prevent disease and provide primary health care can result in many debilitating and easily prevented diseases becoming endemic. These diseases include diarrhoea, dysentery typhoid, intestinal worms and food poisoning. Many health problems are linked to water, its available quality, the ease in which it can be obtained and the provisions taken for its removal after use. The results revealed that the majority of the respondents fetch their water from canals contaminated by the irrigation subsurface return flows responsible for several waterborne diseases.

Table 2.1: *Main environmental problems according to household interviews*

	Mwea Irrigation Scheme		West Kano Scheme	
Problem identified	Freq. N-324	%	Freq. n-55	%
Overcrowding and congestion	207	63.8	46	83.6
Indoor pollution	97	29.9	39	70.9
Water Pollution	186	57.4	43	78.1
Lack of proper latrines	200	61.7	50	90.9
Lack of adequate energy resources	114	35.1	31	56.3
Improper solid waste disposal	87	26.8	41	74.5
Lack of domestic water supply	300	92.6	54	98.1
Lack of privacy	226	69.7	49	89.1

Source: Field survey

The major sources of energy mentioned by the respondents were firewood, paraffin, charcoal, animal dung, and crop residues. The majority of the respondents interviewed mentioned firewood as the only main source of energy while others combined it with other sources. It was observed that firewood used in the villages studied was either bought from a local market or from the surrounding off-scheme reserves. Informal interviews with the firewood and charcoal sellers revealed that dealers outside the scheme supplied these energy sources. Attempts were made to enquire what the approximate daily firewood requirements were. The average daily household requirement for firewood was three bundles of 15 kilograms of wood each. It is noted here that such high demand of wood fuel may lead to deforestation outside the scheme. All the respondents used kerosene for lighting fires and only a minority used it for cooking. A minority of the respondents used charcoal for cooking.

As revealed by the results of the interview survey, animal dung was used in almost all the villages studied. This was because animal dung serves both as a source of energy and a repellent of mosquitoes in the houses.

It was observed, however, that there was a general shortage of wood fuel in the schemes and most of the respondents saw this as a major problem. Another major problem area revealed by the survey was indoor pollution. This was caused by smoke from open fires using firewood, animal dung, crop residues and other biomass fuels. Relatively inefficient lamps used for indoor lighting were also a source of smoke and fumes, which could be dangerous. As seen earlier, the small housing units necessitated the use of one room for kitchen, storage and bedroom. Many respondents were aware that smoke and fumes from wood, animal dung, crop residues, and other biomass fuels as well as smoke from ineffective and inefficient lamps could lead to serious respiratory problems.

Major sources of energy used in the study area are firewood, animal dung and crop residues. Open fires used indoors for cooking produce fumes and smoke, which can cause or contribute to serious respiratory problems. Another problem identified in the study areas was the fact that these energy resources were not readily available. Omissions at the planning stage of the tenants' energy requirements have lead to the depletion of the few bushes in the settlement. Forested areas in the hinterland of the scheme have even been further depleted. The results show that supplies from outside the scheme meet much of the fuel wood requirements and that this will become a looming problem if the trend is not reversed.

CONSEQUENCES ON HUMAN HEALTH

Although there is a wide range of material on the global issues of irrigation projects and health, the following are most common problems in the case of Kenya. They are also directly or indirectly related to the factors identified in the previous section of this chapter (table 2.1):

Malaria: The proportion of people bitten by mosquitoes at home is higher in the morning and evening. The frequency of bites increases during the wet rice-growing season (April to December). People use repellents but only a low percentage use mosquito nets, which shows that the consciousness of mosquito control is not high. The majority of the respondents mentioned malaria as one of the common diseases they suffer from in the course of the year. This shows that malaria is endemic at the two irrigation schemes. Secondary information from medical records from clinics and health centres within the schemes supported our field surveys by showing that malaria is endemic and that the number of cases increase after the flooding of the paddy fields.

Schistosomiasis: It was observed that people tend to live near water such as canals, rivers or ponds and use water for domestic affairs and for body hygiene. All the tenants work in the paddy fields without any protection. They are aware that snails are the intermediate host of bilharzia but control was not progressing.

Amoebic dysentery: Many respondents listed amoebic dysentery as one of the more common diseases that they had been suffering from, certainly in the last 6 months. Most of the respondents were of the opinion that dysentery occurs as a result of the poor quality domestic water supply.

Typhoid: In the course of the study period there was an outbreak of typhoid in at least five villages in the settlement. In fact, a local health centre confirmed an outbreak in one of the villages studied, which had claimed at least four lives. Typhoid is transmitted directly by water ingestion. Most respondents noted that the disease was more severe because there no drugs were available in the government aided health centres and expensive private clinics had to be relied on.

Cholera: Cholera was also mentioned as common water borne disease. Official records showed that there had been at least an outbreak of cholera in some villages. Because of poor sanitary systems and congestion in villages added to poor quality drinking water, the spread of cholera was noted to be rapid in the villages where there had been an outbreak.

Others: Other common diseases mentioned were: Hepatitis, which is transmitted through ingestion of contaminated water; bronchial pneumonia and cardiac failure. Meningitis was also mentioned by quite a large number of respondents and finally, intestinal worms. In general, all the respondents noted that these diseases affected their working schedules seriously and ultimately affected final crop yields. The diseases associated with rice growing as well as those identified in this study can for convenience sake are divided into three categories:

(i) Diseases transmitted by vectors or intermediate hosts living in the rice fields or associated irrigation canals i.e. malaria and schistosomiasis,
(ii) Water related diseases such as dysentery contracted through contaminated drinking water i.e. amoebic dysentery, typhoid, cholera, intestinal worms, diarrhoea and food poisoning, and finally,
(iii) Those related to insanitary housing conditions and overcrowding i.e. cholera, tuberculosis, influenza, meningitis and pneumonia since they can be transmitted from one person to the other.

In the study areas, vector borne diseases were seen to have the greatest socio-economic impact on the community. Large numbers of the population are affected causing a high mortality rate especially among infants and adolescents. If the diseases are not directly fatal they have a debilitating effect on chronic sufferers. These diseases reduce working ability of those infected as well as minimizing the quality of life. The most important of these diseases is malaria transmitted by mosquitoes and schistosomiasis, which has various aquatic or amphibious snails as its intermediate hosts. Schistosomiasis is in fact the most prominent and widespread disease after malaria in the study area.

Despite vigorous anti-malarial efforts, malaria continues to be a major disease. Malarial infections are brought about by different malaria parasites of the genus plasmodium of which the most common are *P. foiciparum and P.vivax.* Schistosomiasis ranks among the most important occupational and public health problem in the study areas. Rice cultivation does not invariably lead to a schistosomiasis problem. For example, collaboration between the NIB and the MOH prevented snails from becoming established at the Ahero Pilot Scheme; this action was prompted by the disastrously prevalent rise in schistosomiasis in

Mwea. It was noted that environmental changes due to irrigation development transformed seasonal and moderately prevalent malaria into a currently permanent hyper-endemic situation in the two rice schemes.

The range of economic consequences of ill health is difficult to assess. However, the economic loss through absenteeism caused by schistosomiasis, was estimated at US$ 8,640,000 annually. Extrapolating from studies elsewhere, the production loss due to ill health was estimated at 10%, which in the case of schemes under the NIB means an annual loss of US$ 818,700 (WHO/FAO/UNEP, 1988, quoted in WHO/FAO/UNFP, 1990). Although the present study has not calculated the economic costs of ill health, the results show that farmers in Mwea and West Kano suffer a number of debilitating illnesses which, if the economic loss were estimated, would be enormous. It is therefore observed that to ensure timely considerations of health in the planning of irrigation development, the opportunities for inter-sectoral co-operation have to be utilised at the crucial moment of the project cycle.

Table 2.2: *Common human diseases identified*

	Mwea Scheme		West Kanno Scheme	
Disease	Freq. n=324	%	Freq n=55	%
Malaria only	86	26.5	11	20
Schistosomiasis only	62	19.1	8	14.5
Amoebic dysentery only	18	5.5	4	7.2
Typhoid only	11	3.4	5	9.1
Cholera only	15	4.6	7	12.7
Malaria, Schistosomiasis Amoebic dysentery, Typhoid and cholera	131	40.4	18	32.7
Other diseases	11	3.4	2	3.6
TOTAL	324	100	55	100

Source: Field survey

Malaria, schistosomiasis and other water borne diseases identified have caused great damage since considerations for public health effects have been neglected by the NIB in general and the two schemes' managements in particular (table 2.2).

The interview survey revealed that the major environmental problems perceived by respondents were of two kinds: those that are social in nature and those that are physical.

Environmental problems as perceived by the respondents are human health and poverty problems. Physical environmental problems are increasing due to aquatic weeds, salinity and alkalinity in the fields as well as water logging. It was the feeling of many respondents that the main problem in the social sphere in the settlement was the human health problem. However, the respondents noted that poverty was the most important social environmental problem facing the settlement. This problem is discussed below. The remaining respondents found other social problems important other than the two mentioned above. Among those of interest to this survey were the causes of the problems mentioned above rather than their results.

Poverty is one of the main factors behind the intensity of certain diseases. If poverty is defined as a lack of assets to pay large expenses or to meet contingencies, then powerlessness, malnutrition, high birth rates, high infant mortality, vulnerability to diseases, physical weakness and indebtedness indicate poverty. According to Chambers, et al. (1991:112) in *Rural Development: Putting the Last First,* diseases like cholera are indicators of poverty. Overcrowding added to inadequate water and sanitation facilities create the right conditions for its transmission. This disease seldom strikes those in upper socio-economic groups, even in regions where it is endemic. It was the feeling of most tenants that poverty, or from the tenants' point of view, poor living standards and conditions, is the major cause of ill health in their scheme. During the planning stage of the settlement the planners were preoccupied with the main objective of increasing agricultural productivity and farmers' welfare was peripheral to this goal. The management exercises a high degree of control over tenant activities. The overall system of control is very strict. It is based on irrigation rules (cap 347) and the success of Mwea and West Kano has largely been attributed to the close supervision that has protected tenants from failure.

However, this type of control has made tenants apathetic and they now tend to mistrust the board and see themselves as slaves. The almost complete control over tenant activities, the far reaching sanctions, the absence of legal rights to holdings, or the assurance of the inheritance of the holdings have led to tenants' sense of insecurity. The regime under which the schemes are managed, and to which most of its success is attributed, is probably its most disquieting feature. Much depends upon whether the present situation is considered a transitional phase during which the tenants are trained to be independent. The other option is whether training is a built-in feature of these schemes, without which it would eventually fail - as has happened in other irrigation settlements in Africa. An irrigation scheme that does not empower its participants would inevitably perpetuate peasants' apathy, dependence, lack of initiative, while driving others to search outlets for initiatives outside planned settlement schemes.

MACRO-LEVEL IRRIGATION RELATED ENVIRONMENTAL PROBLEMS

When water does not conform to water quality standards, either existing or intended, water is polluted. Water is contaminated when real health hazards exist or damage is risked from using it for specific purposes such as drinking water. From this field survey and the results of the interview schedule, most respondents used canal water for domestic purposes. Water samples analysed showed an increase in coliform counts, which is an indicator of the presence of disease-causing pathogens in the water. These pathogens eventually drain into the canals used as sources of domestic water supply. An increase in hardness of water was found, indicated by an increase in calcium carbonate ($CaCo_3$), dissolved salts increasing conductivity, showing the surface drain water near exits to be highly polluted. *(I hope I have not misunderstood the previous sentence.)* This was also indicated by an increase in chlorides, nitrates, calcium, magnesium, sodium, potassium and sulphate. When irrigation water finally flows into receiving water bodies the quality of the water is ultimately affected and will then pose danger to aquatic, animal and human life.

Adverse effects of irrigation on water quality are an increase in salinity, turgidity, colour, taste, temperature, nutrients, bacteria and viruses (indicated by increase in coliform count). These effects can cause economic losses and deterioration of the environment (tables 3 and 4). The rise in salts due to irrigation usually has the effect of increasing hardness. Part of the irrigation water flow goes back into the canals and rivers used as sources of drinking water. Ingestion of this contaminated water has led to widespread waterborne diseases in the two schemes studied. The eventual pollution of receiving water bodies is equally important, as it poses real danger to aquatic inhabitants, animals and downstream human habitation.

The migration movement of salts in water, the accumulation through evaporation and the eventual deposition are responsible for the salinity of the soil. Dissolved salts migrate in artesian waters through tectonic cracks and re-migrate to the surface. Salinity may then develop as a result of evaporation during very dry or hot conditions. Secondary salinity often occurs following irrigation as a result of the water table rising above the minimum level.

The concentration of soluble salts in the soil solution increases as water is removed from the soil by evaporation and transpiration. Alkalinity may result from the evaporation of water containing low concentrations of sodium salts. Secondary salinity is a very quick process as compared to alkalinity. These two problems are caused by the accumulation of bases in the rooting layer. As observed in some areas, the white accumulation of bases was obvious in some units. The concentrations sometimes were above critical levels in some sections. The damages caused by accumulation of bases are divided to two types:

- High total concentration of bases of which Calcium (Ca) is a major base, is an indication of i.e. salinity. This causes the inhibition of water uptake by plants because of high osmotic pressure. The measure of conductivity is a way of estimating soil salinity in general. The higher the conductivity, the higher the level of salinity (tables 4 and 5).
- High pH because of sodium (Na) is an indication of alkalinity. Where Na is high, Ph becomes high in fact > 8.5. The alkalinity causes deficiency in some elements of which availability is low in high pH. Na inhibits Potassium uptake. Alkalinity can be proved by pH and sodium concentrations. The accumulation of bases on the surface is caused by poor drainage (tables 4 and 5).

The most important factors causing water logging, salinity and alkalinity in Mwea and West Kano are aridity of the climate together with geomorphology, topography, physiochemical characteristics of the soils and soil and water management practices. The presence of a high water table and high evapo-transpiration rates in the paddy fields in Mwea and West Kano, account for secondary salinity and sometimes alkalinity of soils. The development of water logged soils is mainly associated with low-lying lands with poor quality soil and poor internal drainage. As a result of intensive and continuous application of water in the paddy fields in Mwea and West Kano, with the absence either of efficient drainage systems or of good internal drainage in some units, water logging may

Table 2.3: *Water quality of constituent applied irrigation and surface drain water near exit of irrigation water: irrigation season - Mwea irrigation scheme (average for 10 sampling sites). **

Constituent or characteristic	Applied Water	Surface Drain Near Exit
Temperature ^0C	18.0	21.4
Dissolved Oxygen	10.2	8.0
pH units	7.62	8.5
Hardness as $CaCo_3$	44	168
Turbidity Units	32	146
Conductivity mmhols/cm	80	263
Chlorides as mg/l	0.98	9
Nitrates as mg/l	0.25	6.5
Calcium as mg/l	9.8	33.2
Magnesium as mg/1	4.6	18.7
Sodium as mg/1	4.0	20
Podium as mg/1	1.4	4.6
Sulphates mg/l	3.6	24.6
Total Dissolved Solids (TDS)	121	470
Coliform per 100 ml	150	1,079

Source: Field survey

- Analysis done at the Government Chemist, Nairobi Nov. 18, 1996 and the results explained and discussed with the researcher.

Table 2.4: *Water quality of constituent applied irrigation and surface drain water near exit of irrigation water: irrigation season - West Kano scheme (average for 10 sampling sites)**

Constituent or characteristic	Applied Water	Surface Drain Near Exit
Temperature ^0C	17.9	20.2
Dissolved Oxygen	11.0	8.4
pH units	7.4	8.8
Hardness as CaCO3	58	142
Turbidity Units	42	128
Conductivity mmhols/cm	66	297
Chlorides as mg/l	0.58	11
Nitrates as mg/l	0.08	4.5
Calcium as mg/l	6.65	68.2
Magnesium as mg/l	7.2	22.6
Sodium as mg/1	3.86	34.5
Potassium as mg/1	0.76	12.6
Sulphates mg/l	4.83	44.3
Total Dissolved Solids (TDS)	263	657
Coliform per 100 ml	215	2,175

Source: Field survey

- Analysis done at the Government Chemist, Nairobi Nov. I8, 1996 and the results explained and discussed with the researcher.

2.5: *Soil characteristics in selected areas in Mwea settlement*

Parameter\ Station	K2	W7	H6	M13	T19
pH- H_2O (1:2.5%)	7.94	5.79	8.29	6.68	7.72
Hcl Reaction (10%)	X	X	00	X	00
CEC (ml/100g) Ph 7.0	61.0	68.0	76.0	70.0	68.0
Exch. Ca. (me/100g)	45.5	36.0	136.9	34.2	52.6
Exch Mg. (me/100g)	19.9	15.3	29.9	20.2	31.2
Exch K (me/l00g)	0.5	0.3	0.1	0.1>	1.0
Exch Na. (me/100g)	1.0	0.5	6.8	3.5	3.4
Avail. P ppm	15.3	25.3	24.9	22.7	94.8

Horizon 0-15 cm.
K2 = Karaba Unit K2
W7= Wamumu Unit W7
H6 = Thiba Unit H6
M13= Mwea Unit M13
T19 = Tebere Unit T19.
X = no reaction at all
00 = slight reaction

Source: MIAD Soil Labs Feb.1992.
NB. The units where soil analysis was done with the assistance of the irrigation officers in charge had been selected as problematic. Deterioration of the schemes' soils was noted. Reclamation of the already affected soils will depend among other things on the proper selection of the method best suited to the specific conditions in the field. In order to reach such a decision, field and laboratory work is required to indicate the extent of the problem.

Table 2.6: *Soil characteristics in selected areas of West Kano Scheme*

Parameter/ Station	**WK.1**	**WK.2**	**WK.3**	**WK.4**	**WK.5**
PH H_20 (1:2.5%)	7.65	7.25	6.30	6.45	7.45
Hc1Reaction (10% Hc1)	X	00	X	X	X
CEC (me/100g) Ph 7.0	68.5	65.4	75.2	69.3	66.7
Exch. Ca. (me/100g)	52.2	38.5	70.6	41.5	31.2
Exch. Mg. (me/100g)	20.6	14.9	2.6	24.8	28.8
Exch. K. (me/100g)	0.4	0.1	0.1	0.6	0.9
Exch Na. (me/100g)	2.0	0.8	0.8	1.0	5.4
Avail. P. ppm	70	34.5	30	18.9	76

Horizon 0-15 cm.
WK. I - West Kano station 1
WK.2 - West Kano station 2
WK.3 - West Kano station 3
WK.4 - West Kano station 4
WK.5 - West Kano station 5
Source: JICA 1992.
N.B. This analysis was done during a feasibility study by the Japanese International Co-operation Agency in 1992.

become a serious problem in the scheme. Quantitative data on the extent of the present and potential salinity, alkalinity and water logging was not available whereas this should be a research priority. However, from the examples given in the results, it should be noted that these problems, although of low magnitude, should be mitigated in order to avoid future damage to human health.

INSTITUTIONAL ISSUES

The complexity of the hydrological, geological, chemical, and ecological factors presented in chapter four poses a tremendous management challenge. They are joined and often exacerbated by an equally complex and conflicting maze of political, social, economic, legal, administrative, and other institutional factors. The technical complexities affect the proposed problem and awareness planning initiatives. The institutional complexities affect problem awareness and the acceptance and effectiveness of the proposed planning methodology.

It is typical that for institutional reasons some changes in the environment are judged to be a problem rather than a merely a change. These institutional factors will then influence the choice of a response. Because institutional and scientific considerations are entwined, responses must be based on an understanding of the complex interactions that can occur between these two dimensions.

The institutional barriers often are more difficult to overcome than the technical problems. In addition, the separation and fragmentation of institutional factors add to the problem. The technical complexities affect the proposed problem and planning initiates. The institutional interests and responsibilities make it hard to develop a coordinated strategy for integrating environmental and health considerations in irrigation development planning.

Institutionalization also includes administrative organization, social customs, regulations, policies and laws. Institutional analysis must consider economic arrangements such as pricing policies and constraints, subsidies, water marketing and allocation as well as combining surface and ground water management. Social acceptability has proved to be an increasingly visible criterion to address environmental and health concerns. It is particularly important to recognize the fundamental role that social issues play in setting the stage for both the creation of environmental and health problems and the choice of solutions.

Another critical area to be considered when integrating environmental and health considerations is the influence of agencies and organizations, including both governmental and non-governmental bodies. Organizational influences change over time and reflect nations' changing priorities and values. Questions of environmental degradation and problems related to human health generally were initially rarely considered because society did not give these issues the high priority it does today.

The legal issues involved in integrating environmental and health considerations in irrigation development are complex and they are closely related to the organizational issues. Environmental Impact Assessment has been suggested as an efficient environmental planning tool in irrigation development. However, legal mechanisms have to be in place to ensure that EIA is done before any irrigation project is started. This ensures that adverse environmental and health effects are addressed at the planning stage so that appropriate measures can be taken beforehand (table 2.7).

2.7: ***Planning and management strategies at the household level***

Problem identified	Planning and management strategies and percentage of respondents
Housing conditions and congestion at home	A small minority (36%) of the respondents have bought land outside the scheme and have settled some of the family members there.
Water supply and sanitation	Use of chemicals like alum to treat water (22%), boiling drinking water (49%), community participation in the construction and drilling of bore holes (10%), use of several containers to settle the total suspended solids of canal water (86%).
Indoor pollution	Majority of respondents said they do nothing about this problem (98%). Only 2% said they dried the firewood before use to reduce smoke
Energy sources problems	Majority of the respondents have embarked on tree planting in the villages to counter this problem (78%) Use of crop residuals (45%) and animal dung (60%) was also a strategy used to cope with energy shortage problems.
Treatment of common human diseases	Chemotherapy (96%) traditional herbs (46%) repellents (76%), Mosquito nets (37%) protective clothing (0%), engine oil used as repellent for snails (48%), Community participation in building dispensaries and health centres
Increased aquatic weeds	Early and continuous weeding
Salinity problems	Use of excess fresh water
Poverty and poor living conditions	All the tenants interviewed have joined Rice Farmers Amalgamated Co-operative Savings and Credit Societies
Water-logging	Improve drainage and raising of budding

Source: Field Survey

Planning and management strategies at the scheme level are summarised in 2.8 below. These results were generated by key information interviews held with management staff, local leaders and community development officials at the divisional level. The District Development Plans (1994/96) stipulate what the measures are to be during the planning period which includes, among other things, the construction of dispensaries and health centres to cope with the high incidence of diseases in the schemes.

In 1991, the African Biodiversity Institute (ABI) initiated a project in Karaba Section of Mwea which integrates environmental, health and development issues through the provision of water, sanitation and drugs. Its priority is the elimination of common human diseases at Mwea. From the time the project was started - with funding from the UNICEF, the UNDP, the Africa 2,000 Network, and the Government of Japan - it has constructed wells, pit latrines and laundry units and has provided drugs and mosquito nets to the population. Six water pumps and four latrines have been built. Three more projects are being constructed by ABI and NIB in Kiandegwa village in

Mwea. This initiative takes into account the role of community participation in environmental management.

Table 2.8: *Planning and management strategies at the scheme level*

Problem	Planning and Management Strategies
Housing conditions and congestion	NIB give material loans to farmers to construct houses; extension of some congested villages.
Water supply and sanitation	Community health and personal hygiene education.
Canal water quality	Appropriate water management strategies whereby all the applied water is utilised by the rice crop.
Energy sources problems	Encouraging afforestation in the villages, encouraging research on the use of rice husks to make charcoal
Common human diseases	Participation in the construction of health centres in the scheme; the purchase of drugs for the staff; community education; encouraging and corroborating with NGO's working with these communities, used to spraying houses and mollusciciding *(this word is not in the dictionary)* in the canals.
Increased aquatic weeds	Encouraging early and frequent weeding
Salinity problems	Frequent monitoring of salt levels in the soils; field days with tenants to identify problem areas; improvement of drainage.
Water-logging	Improvement of drainage, lining of canals and reclaiming abandoned water-logged plots by JICA.

Source: Field survey

CONCLUSION

Irrigation development leads to specific problems unique to that system. These problems have not been met with the appropriate planning and management strategies more so because attempts to understand the complex factors interacting in the system has not been understood by neither the schemes management nor the NIB in general.

The establishment of these irrigation schemes has led to great modifications in the environment giving rise to the spread and multiplication of disease vectors and contributing to increasing human health problems. These developments - the considerable environmental and socio-economic changes that have taken place - demand an integrated health control strategy. Irrigation development brings greater numbers of people into closer contact with, for instance, water-borne diseases. There is, therefore, a pressing need for careful and unbiased appraisal to identify irrigation development approaches and management strategies that can reduce social costs and deliver benefits with the lowest possible environmental damage. More care is needed to assess in advance the environmental and socio- economic impact of irrigation development.

The main purpose of an irrigation scheme is to improve the well being of the population. If a well-planned research policy is not drawn up at the initial stages of irrigation development, this purpose is defeated. It has been seen that emphasis in these NIB schemes has been to increase productivity whereas little has been done to maintain the health of the farmers. Because of this unforeseen fact in irrigation planning human health continues to decline to the detriment of the eventual production of irrigation development.

Irrigation development involves not only supplying agricultural land with adequate water but also requires from the very beginning that measures be put in place to control, contain and minimise environmental health problems. Careful considerations should therefore be given in the planning of irrigation projects, to conduct thorough investigations of factors such as natural drainage, geology, hydrogeology, soils, salts present salinity and disease vectors that are likely to have an impact on human health.

While the benefits of irrigation are widely recognised, any arising environmental and health problems must also be paid attention to and programmes should be mounted in the planning stage to assess environmental and health hazards. However, it has been noted that no legislation exists that enforces this.

REFERENCES

Aral, A., (1976) "Irrigation and Drainage in Relation to Salinity and Water-logging". In: Salinity *Seminar, Baghdad*, pp. 56-119.

Arnon, I. (1981*) Modernisation of Agriculture in Developing Countries: Resources, Potentials and Problems*. John Wiley and Sons U.S.A. p. 79-132.

Biswas, A. K. (1984*) Monitoring and Evaluation of Irrigated System.* International Journal of Water Resources Development 2 (1), pp.3-25

Biswas, A. K. (1987) *Monitoring and Evaluation of Irrigated Agriculture: A Case Study of Bhima Project*, India, Food Policy February.

Biswas, A. K. (ed.) (1994) *International Journal of Water Resources Development: Irrigation Development in Asia*. Vol. 10(4).

Biswas, A. K. (1994) *Considerations for Sustainable Irrigation Development in Asia*. In: *International Journal of Water Resources Development*. Vol. 10(4).

Bottrall, A. (1978) "The Management and Operation of Irrigation Schemes in Less Developed Areas", in Widstrand, C. (ed.) *The Social and Ecological Effects of Water Development Projects in Developing Countries*. Oxford: Pergamon Press.

Carward, E. W. (1991) "Planning Technical and Social Change in Irrigated Areas. In*: Putting People First: Sociological Variables in Rural Development* edited by Cernea, M.M. Washington, D.C: A World Bank Publication. Oxford University Press.

Chambers, R and Jon, M., (1973) *Mwea – An Irrigated Rice Settlement in Kenya*, Munchen: Weltform – Verlag.

Chancellor-Weale, F. M. (1991) "Smallholder Irrigation Design Features and Their Long Terms Effects on the Immediate Environment-examples From Zimbabwe, in Wooldridge, R. (ed.), *Techniques for Environmentally Sound Water Resources Development*. London: Pentech Press, pp. 106-117.

Currunthers, I. D. (1968) *Irrigation Development Planning-Aspects of Pakistan Experience*, Department of Economics, Wye College-University of London.

Currunthers, I. D. (1978) "Contentious Issues in Planning Irrigation Schemes. In: *The Social and Ecological Effects of Water Development Projects in Developing Countries* e, in Widstrand, C (ed.). Oxford: Pergamon Press.

Currunthers, I. D., and Collins, C. (1981) *The Economics of Irrigation*, Liverpool: Liverpool University Press. Press.

FAO, (1971) *Salinity Seminar, Baghdad*. Irrigation and Drainage Paper. Report of Regional Seminar on Methods of Amelioration of Saline and Water-logged Soils. Baghdad, Iraq 5-14th December 1970.

FAO/UNESCO (1973) *Irrigation, Drainage and Salinity*. An International Source Book. Hutchiston and Co. Ltd.

FAO (1986) *Irrigation in Africa South of Sahara*. Investment Centre Technical Paper No. 5. Rome: Food and Agriculture Organization.

FAO (1987) *Effects of Agricultural Development on Water-borne Diseases*. Edited Versions of Working Papers Presented to the 7th Annual Meeting of the Joint WHO/FAO/UNEP Panel of Experts on Environmental Management for Vector Control. 7-11 September 1987. Rome: Food and Agriculture Organization.

GOK (1979) The National Master Water Plan. Nairobi: Kenya: Government Printer.

GOK (1993) Kenya Demographic and Health Survey. Government Nairobi: Kenya: Printer.

GOK (1994) National Development Plan 1994-1996. Nairobi: Kenya: Government Printer.

Hardoy, J. E., Diana, M, and David, S., (1993) *Environmental Problems in Third World Cities*. London: Earthscan Publications Ltd.

Hotes, F. L. and Erman A. Pearson (1977) "Effects of Irrigation on Water Quality". In: *Arid Land Irrigation in Developing Countries. Environmental Problems and Effects*, in Worthington E. B., (ed.) Oxford: Pergamon Press.

Ireri, D. (1986) *Mwea Tebere Irrigation Scheme Lessons and Experience*. In: Large-Scale Irrigation Development in Kenya. Past Performance and Future Projects, in Ruigu G. M. (ed.) Report No. 23/1988. Leiden: ASC/Food and Nutrition Studies Programme.

Japanese International Co-operation Agency (JICA) (1988), "Feasibility Study on Mwea Irrigation and Agricultural Development". Main Report. JICA.

Japanese International Co-operation Agency (JICA) (1992), "Feasibility Study of Kano Plains Irrigation Project". JICA, 1992.

Kovda, V. A. (1980) *Problems of Combating Salinization of Irrigated Soils*. (Selected Lectures) Moscow.

Modhar, H. (1993) "Report on Mwea Irrigation Settlement-Layout and Growth". A Pamphlet Issued to Visitors on Field Trip.

Myers, L. E. (1984) "Environmental Management for Vector Control in Rice Fields in the Context of an Integrated Approach", in *Environmental Management for Vector Control in Rice Fields*. Marthar, T.H. *et al.* FAO, Irrigation and Drainage. No. 41, pp. 124-129.

National Irrigation Board (NIB), Annual Reports on MIS for 1987 1988, 1989, 1990, 1991, 1992, and 1993.

Odero A. N. (1992) "A Socio-economic Study of Some Performance Aspects of West Kano Irrigation Schemes, Kisumu District". Unpublished Mphil Thesis, Moi University. p.1-33.

Pazvakavamba, S. (1987) "The Impacts of Irrigation Development on the Environment: A Policy Perspective". Paper Presented in a Seminar on Irrigation Farming in Kenya and Zimbabwe, in Ruigu, G. M. and Mandivamba, R., (eds.), *Irrigation Policy in Kenya and Zimbabwe*. Nairobi: Institute of Development Studies, University of Nairobi.

Rajappa, R., P. Krishna, A. Sharma and A. K. Sharma. (1993) "Environmental Impact Assessment for Subarnarekha Irrigation Project". In: International Journal of Water Resources Development. Vol. 10, No. 2.

Roa, D. O. (1987) *Rural Development Through Irrigation*. New Delhi: Ashish Publishing House.

Reddy, M. V. (1991) "Environmentally Sound Irrigation System Development: Problems and Prospects- a Case Study from India", in Wooldridge, R. (ed.), *Techniques of Environmentally Sound Water Resources Development* London: Pentech Press, pp. 211-228.

Ruigu, G. M., (1987) *Large Scale Irrigation Development in Kenya: Past Performance and Future Prospects*. Report No. 23 – Food & Nutrition Planning Unit, Ministry of Planning and National Development, Nairobi & African Studies Centre, Leiden, Netherlands.

Singh, P. P. and A. A. Afroz (1985) "Environmental Impact Assessment of Saryu Canal Irrigation Project of Eastern U. P. India". In: *Environmental Impact Assessment of Water Resources Projects*. Proceedings of the International seminar held at Water Resources Development Training College, University of Roorke, University Press, pp. 49-55.

Sharkar, V. and Suresh Kurmar (1993) "Of Arid land Irrigation and Some Combating Measures". In: *Arid Land Irrigation and Ecological Management* edited by Singh, S. D. Scientific Publishers, Jodhpur, India.

Thitai, M. W. (1990) "Planning Procedures for Water Resources Development. In: Seminar on *Water Resources Development and Vector-Borne Diseases in Kenya*. WHO/FAO/UNEP. PEEM Secretariat Kisumu.

White F. G. (1978) *Environmental Effects of Arid Land Irrigation in Developing Countries*. UNESCO. Prepared in co-operation with UNEP and SCOPE. MAB Technical Notes No. 8.

WHO (1965) *Snail Control in the Prevention of Bilharziasis*. WHO Monograph Series No. 50. Geneva.

WHO (1990) *Epidemiology and Control of Schistosomiasis* Report of a WHO Expert Committee. WHO Tech. Report Series No. 643. WHO Geneva, pp. 45-4.

WHO/FAO/UNEP (1988) *Seminar on Water Resources Development and Vector-Borne Diseases in Kenya*. PEEM Secretariat, Kisumu, September, pp. 11-13.

WHO, (1993) *The Control of Schistosomiasis* 2nd Report of the WHO Expert Committee. Geneva: WHO Technical Report Series No. 830.

3 Choice and Use of Rural Water Supply Systems:
The Environmental and Socio-Cultural Dimension

Wilson Nyaoro

Many of the rural water supply development projects designed by planners, government agencies and non-governmental organisations remain alien to the people intended to benefit from them because their ideas cannot take root in a new cultural landscape. Technicians normally blame project failures on social institutions, cultural norms, values, attitudes and practices, not knowing that cultural self-image and identity usually resist superimposed change. The result is that a significant proportion of water projects have either been abandoned, vandalised or only function intermittently and have therefore failed to meet the expectations of the intended beneficiaries, governments and donors. Any meaningful and sustainable development of rural water supplies must therefore address the inherent environmental socio-cultural set-up of the target group.

The main objective of this chapter is to use an interdisciplinary approach in order to examine the environmental and socio-cultural factors that influence the development, choice and use of rural water supply systems in Siaya District. To do this effectively, this chapter will identify and critically assess the socio-cultural beliefs and practices, the manner of water consumption (consumptive *this word gives the impression of being associated with tuberculosis)* water-use) and the attitudinal variables (ability and willingness to pay for improved services) in relation to the development, the choice and the use of rural water supply systems. Furthermore, the nature and extent of community participation in the various stages of the project cycle will be examined. The feasibility, flexibility and acceptability of current rural water supply development institutions and the technology used will be assessed according to the terms of local environmental conditions.

Results of this survey have indicated that there was either minimal community participation or none at all during the conceptualisation, planning, design, implementation, operation, administration, maintenance and management of water supply projects. Consequently, socio-cultural beliefs and practices, water-use practices and the local attitudinal variables were never integrated in the water planning process and the existing institutions and the technology used are inappropriate since their design did not take complex environmental socio-cultural variables into consideration. The existing rural water supply development strategies are structurally elitist and debilitating. Lopsided priorities, faulty technological choices and the adoption of inapplicable and irrelevant strategies have contributed to

M.A.M. Salih (ed.), Local Environmental Change and Society in Africa, 2nd Edition, 41–68.

the problems confronting water development projects. Arising from the realisation of the inadequacies of these strategies, it has been recommended that an integrated approach be adopted as an alternative water planning model. In addition to other environmental sub-systems, such a model must consider environmental socio-cultural variables as an integral component of the water planning process. Such a dimension is important because it promotes water supply development strategies rooted in local values and traditional water resource management systems. It is essentially based on meeting people's substantial, procedural and psychological needs and takes into consideration their preferences, their creativity and their active involvement by unleashing local initiatives and capacities.

This chapter strives to examine the sources of disappointments of water development projects in the past with the aim of making water policies with more effective solutions. It focuses on the examination of environmental socio-cultural factors influencing the development, the choices and use of rural water supply systems in Siaya District, Kenya. This chapter pays special attention to institutional arrangements and community participation; to traditional beliefs, practices and attitudes. It examines the appropriateness of technology taking local resource and technical know-how availability into account.

ENVIRONMENT AND LAND USE

Siaya District has a quasi-equatorial climate locally relieved by the strong climatic influence of the expansive Lake Victoria. The amount and the distribution of rainfall in the district are highly influenced by the altitude and the direction of the wind. The westerly winds originating from Lake Victoria, converge with the prevailing easterly winds above the topographically higher parts of the district, causing a strong increase in rainfall. Hence rainfall is closely related to the altitude of the area. The southern parts of the district near the shores of the lake where the altitude is low is characterised by dry and semi-arid climatic conditions getting progressively gets wetter towards the eastern and northern parts where the altitude increases again. These high altitude areas receive mainly conventional and relief rainfall to the amount of about 2000mm annually. In the low altitude southern and western parts, especially near the shores of Lake Victoria, only about 700mm of rainfall is received annually. Rain is received in the long wet season, which run from February through May and in the short season, which lasts from October to December.

Evaporation decreases with an increasing altitude, resulting in an average evaporation of 1800-200 (mm/year) in the southern parts to about 1600 – 1800 (mm/year) in the north-eastern parts of the district. Temperatures in the lower (southern) parts of the district vary from a mean minimum of 17°C to a mean maximum of 30°C, with an annual mean temperature of 23°C. In the higher altitude parts in the north-east of the district the temperature is a few degrees lower resulting in a mean annual temperature of about 21°C.

The most important food crops grown in Siaya District are maize, cassava, sorghum, finger millet, sweet potatoes, beans, groundnuts and vegetables. In addition, the district is blessed with wetlands such as Yala Swamp which form "high potential" areas for rice growing. The main cash crops are sugar cane, cotton and coffee. Currently, sugar cane growing is mainly concentrated in the eastern parts of Ukwala and the northern part of Yala Division. Coffee is mainly grown in Yala,

Boro and Ukwala Divisions while cotton is grown in the southern part of the district especially in Rarieda and Madiany Divisions. (with the district, especially in Rarieda and Madiany Divisions, the district has a high potential for growing fruits and vegetables but this requires that irrigation facilities be provided (table 3.1).

Table 3.1: *Land use Types Covering Siaya District*

Land use type	Percentage (%)	Cover (KM^2)
Grazing	34.5	870
Fallow	19.2	484
Maize	12.3	310
Bushland	8.9	225
Hedges	4.9	124
Cassava	4.7	118
Bare ground	2.2	56
Structures	1.6	40
Cotton	1.6	40
Roads	1.5	38
Sorghum	1.2	30
Sugar Cane	0.9	23
Woodlots	0.3	8
Others (Finger Millet, Bananas, Sweet Potatoes, Groundnuts, and Vegetables)	1.2	30

Source: Kenya, 1988.

Livestock husbandry is an important activity especially in Boro, Usigu, Bondo, Rarieda and Madiany Divisions. However, most livestock are of the indigenous variety (zibu) and are raised under traditional grazing systems. In terms of forestry, Siyaya district is one of the least forested districts in the country, with only 0.3% of the area forested. With the rapidly increasing human population and the associated activities most of the forested areas have been encroached upon and interfered with, resulting in the disruption of the hydrological cycle.

Siyaya District's potential for both agricultural and livestock development is great, but a combination of low moisture retention capacity, relatively poor soils and unreliability of water supply sources act as major constraints to the development efforts in these sectors. The next section of this chapter examines the demographic profiles of Siaya District and the manner in which they form the relationship between local environment and water management practices.

DEMOGRAPHIC PROFILES AND SURVEY

The results of the 1989 population census records the population of Siaya district at 639, 439 (294, 313 males and 345, 126 females), with an annual growth rate of 3.1%. Using the 1979-1989 intercensal growth rate of 3.1% the population of Siaya District was estimated to increase to 749,071 in 1994 and to 796,674 by 1996 (District Development Plan, 1993-1996). The population increase has put an added strain on the existing limited water resources. A clear relationship between rainfall pattern and population density can be observed. Low rainfall areas in the southern and southwestern parts of the district have relatively low densities of about 200-250 persons per square kilometre, while the northern and north-eastern parts, the areas with the highest rainfall, have densities of about 350-400 per square kilometre (Kenya, 1988). The central district has a moderate population density, with some pockets of high population concentration.

The general health situation in Siaya district is among the worst in the whole of Nyanza Province. This is illustrated by the average life expectancy figure of 38.5 years in the district, which is the lowest in the entire Nyanza Province. This low life expectancy can be attributed to the prevalence of certain diseases illustrated in figure 6. From this figure it is evident that water related diseases are the most frequently occurring (48%) followed by respiratory (13%) and digestive (12%) diseases (Kenya, 1988). Water related diseases include, *inter alia*, cholera, malaria, schistosomiasis, typhoid fever, gastro-enteritis and dysentery. This undesirable situation of disease prevalence has resulted in infant as well as child mortality rates in the district (211 per 1000 and 247 per 1000 respectively) ranging among the highest in Kenya. Hence in Siaya district the need for potable water supply is evident considering that more than 60% of the infant mortality is caused by diarrhoea (Kenya, 1988).

This study was based on an extensive review of rural water supply development policies and investment project reports; published and unpublished literature; and discussions with the relevant people at the local, regional and national levels. Keeping in mind the wide geographic and economic variations in the district, it was thought appropriate to take the estimated 639,439 people (1989 population census) resident in the district as the 'true population.' However, very often the true population is not amenable to sampling in the time available, and a surrogate has to be found. From this point of view it is desirable to develop a sampling frame (i.e. a statistical sub-population from which to take the sample) which locates individuals within the population. It is the 38 clusters of households including Siaya and Bondo rural centres of Siaya District (as pre-determined by the Central Bureau of Statistics, and the Ministry of Planning and National Development) which constituted the sampling frame from which a sample of 380 households (10 households in each cluster) were taken. Therefore the basic sampling unit was individual households (with heads of households as respondents), although in some cases data constraints forced the research team to work with aggregated census information at the enumeration district level.

Out of a sampling frame of 34 rural water supply development projects, the study has focused only on the 17 Existing Water Supplies (W/S) and 4 Under Construction Water Supplies (W/S) giving a total of 21 water supply projects. According to the results of the pre-survey done in October 1995, the other 13 projects categorised as Planned Water Supplies (W/S) are still at the paper work stage and therefore have not been implemented. Water supply project managers, assistant managers, caretakers, and village representatives were targeted as respondents during the survey.

Institutions (both governmental and non-governmental) concerned with rural water supply development in the district were visited. These include Lake Basin Development Authority (LBDA); Swedish International Development Authority (SIDA); Kenya-Finland Cooperation Agency (Kefinco); churches; local project committees; and the Ministries of Water Development, Agriculture, and Environment and Natural Resources. During the institutional survey, the focus was on water planners, physical planners, environmentalists, hydrologists, hydrogeologists, meteorologists, managers, project coordinators, opinion leaders and members of District Development Committees (DDCs) as the respondents at the grass-root or village level.

An intensive study was undertaken of 38 clusters of households, including Bondo and Siaya "rural centres", and 21 rural water supply projects/programmes. To do this effectively, background data was firstly gathered from district water personnel and other personnel from various relevant governmental and non-governmental organisation departments (targeting as respondents the heads of departments or their deputies). Secondly, more specific local information from the water operators and managers of the existing schemes was collected. Thirdly, opinions, attitudes, experiences and other economic and socio-cultural characteristics of the rural dwellers was collected, using the services of research assistants who were first given a short try-out and some training during the initial pilot study phase.

Observations and interviews were conducted with female heads of households (because the women interact most with water supply sources) and/or their husbands, by visiting them both where they work and where they live. Since all households in each cluster were already assigned three digit numbers (i.e. 001, 002, 003 etc.) by the Central bureau of Statistics (CBS) personnel, a randomly selected Random Start (r) and the computed Sample Interval (s) were used to allocate 10 households in each cluster. The research team were then instructed to call only at the (r+s)th. (r+2s)th, (r+3s)th, (r+4s)th, etc. households as they walked systematically around each cluster.

The collection of data from the villages, government and non-governmental organisations (NGOs) was also done through the use of well-designed and structured questionnaire schedules. To avoid getting wrong or irrelevant information (due to misunderstanding/ misinterpretation of the questions) and loss of questionnaires, it was decided that most of the questionnaires were to be handed over to the selected sample population. Only in extreme cases of inaccessibility were the questionnaires posted.

It should be reiterated here that most of the questions asked in the questionnaires were deliberately kept open-ended to enable the respondents to give answers that accurately represent their opinions, values, needs and feelings about the current (and possible future) water supply situation in the study area. However, to ensure that the information obtained by the use of questionnaires was sufficient, emphasis was put on holding meetings and informal discussions. For instance, the possibility of families financing their own water supply projects was discussed and the families' reactions and opinions were listened to. In order to emphasise variations of magnitude and/or associations of the various parameters under observation-examination, various techniques and methods were used in data analysis and presentation. They include descriptive and qualitative techniques.

WATER SUPPLY SOURCES AND THE ASSOCIATED WATER CONSUMPTION PRACTICES

The coverage of the water supply in Siaya District is low on average and unevenly distributed. The northern and eastern parts have a higher coverage, compared to the southern and western parts of the district. Access to safe and clean drinking water depends on the availability and the distribution of the water resources. Field surveys have indicated that the ground water potential is quite limited, particularly in the southern region comprising Madiany, Rarieda, Bondo and Usigu Divisions. Apart from the existing functional piped water supply facilities, the local population depends on river water, lake water, wells/boreholes, dams, ground catchments, springs and roof catchments for domestic needs. In the northern region, the most common potable water supply facilities found are piped systems and wells fitted with hand-pumps. About 50% of the households in Siaya District have access to protected and safe water sources during the wet season and 33% during the dry season. However, it should be noted that these district averages conceal significant variations between and within the divisions.

The divisions in the southern zone such as Bondo, Rarieda and Madiany have very limited access to safe drinking water during the wet season as indicated by the average figures of 19%, 25% and 34% respectively. The situation is worse during the dry season when access to safe drinking water in these divisions is reduced to 16%, 3.1% and 1% respectively. The safe drinking water in these divisions comes mainly from roof catchment and very little comes from the few existing piped water supply facilities usually characterised by an intermittent supply.

The divisions in the northern parts such as Uranga and Ukwala enjoy better access to safe drinking water during the wet season as indicated by the average accessibility figures of 75% and 68% respectively. Based on the safe drinking water coverage figures, the divisions in Siaya district can be ranked in terms of need and priority with Madiany, Bondo and Rarieda as the most needy. The others in descending order of priority are Boro, Yala, Wagai, Ugunja, Ukwala and Uranga.

WATER COLLECTION, STORAGE AND USE

The survey sought to find out by whom water is collected, what the mode of transportation from the source to the household is, and what the distances involved in fetching water from the main source to the household are. These are related to the time spent in fetching water, the type and nature of the storage facilities and the various uses of collected and stored water.

Traditionally, water collection is the domain of women in Siaya district. Women are the chief drawers of water and males are rarely involved in water collection. In the division of labour, men are not associated with this daily routine and arduous task. Table 3.2 shows the distribution of different methods used to fetch water in the divisions. The overall situation indicates that about 90% of the respondents in the district use human porters as their mode of transporting water from the source to the household. Bicycles, handcarts, wheelbarrows and vehicles are not the dominant modes of water transportation. The human porters referred to are the women who carry water on their heads.

Table 3.2: *Percentage distribution of methods used to fetch water by division*

Method	Wet	Dry
Human porter	91.7	80.5
Handcart/Wheelbarrow	1.1	8.2
Bicycle	1.7	1.7
Vehicle	0.1	0.6
Domestic animals	5.3	10.6
Others	0.1	0.1

Source: Field Data

The survey data reveals that 87% of the male adults do not collect water, as opposed to 89% for the female adults who do. There is no gender balance in water collection whatsoever. Even male children appear to be less burdened by this task than female children. Field surveys indicate that 65% of the male children are not involved in water collection as compared to 56% of the female children. Men are mostly concerned with the water once it reaches the house but not how it gets there. In the few cases when men were involved in water collection, they would do so only when the females were sick and they would use mainly domestic animals (e.g donkeys), wheelbarrows, vehicles and bicycles.

One of the main interests of this study was to find out what the distances were that members of households had to travel to fetch water both during the wet and dry seasons. However 87.6% and 65.7% of the households in the district draw their water at a distance of less than 1 km during both the wet and the dry seasons. Survey results indicate that in the dry season a higher number of households fetch water from sources beyond 1 km. Bondo is the most disadvantaged division with 64.3% of the households drawing water from sources beyond 1km during the dry season, followed by Rarieda (63.4%) and Madiany(54.1%). According to the results shown in table 3.3, Madiany division has the highest proportion of households (32.8%) who draw water from distances of between 3 and 4 kilometres.

In addition to distances to water sources, the survey also sought to find out the number of hours taken by members of households to fetch water from the main sources. As shown in table 3.4, in Bondo division only 32.7% of the household members spend less than one hour to get water during the dry season, followed by Rarieda division with an average of 35.0%. This shows that most of the people in the two divisions get their supply of water at distant places which therefore takes more than one hour during the dry season. In Madiany division, 33% take 3-4 hours to fetch water from the main water sources during the dry season while 5% take 5-8 hours to fetch water.

However, tables 4 and 5 should be interpreted with caution. Field surveys sought information on distances and time taken to the main water sources. In most cases these sources are rivers, lakes, dams, ground catchments and unprotected springs and wells. The figures given in the tables give the impression that people have access to water sources within easy reach. This is true in terms of any water source but not necessarily true in terms of safe and clean water sources.

Field surveys revealed that clay pots are the most commonly used containers for storing household. They are used by 93% of the households in Ukwala, 87.5% in Uranga, 85% in Wagai and 83% in Boro. On average 65% of the households in According to the results from the field survey, the most common use of water in the household is cooking, drinking and washing which together account for 98.2% of the total water use. Livestock watering is hardly done at the household level except for a small number of livestock and households with zero grazing grade cattle. Table 3.4 provides data on the distribution percentage of household water use in the district. Siaya District use clay pots for water storage, 12.7% of the households use plastic buckets for storing water while only 10.3% of the households use jerricans for the same purpose. Metal drums and sufurias are rarely used for water storage (table 3.5).

Table 3.3: *Percentage distribution of distance to main water source from the household by division and season*

Division	Less than one km		1-2 kms		3-4 kms		More than four kms	
	Wet	Dry	Wet	Dry	Wet	Dry	Wet	Dry
Yala	95.8	95.8	4.2	4.2	-	-	-	-
Wagai	80.0	77.5	20.0	22.5	-	-	-	-
Ukwala	86.9	67.2	13.1	27.9	-	4.9	-	-
Ugunja	96.9	97.0	3.1	3.0	-	-	-	-
Boro	61.7	55.3	36.9	42.6	1.4	2.1	-	-
Uranga	85.0	80.0	12.5	12.5	25.5	7.5	-	-
Bondo/Usigu	100.0	67.7	-	43.9	-	19.4	-	1.0
Rarienda	100.0	36.6	-	53.4	-	10.0	-	-
Madiany	82.0	45.9	18.0	21.3	-	32.8	-	-
District	87.6	65.7	12.0	25.7	0.4	8.5	-	0.1

Source: Field Data

Table 3.4:*Percentage distribution of time spent to fetch water in hours by division*

Division	Less than 1 hour		1-2 hours		3-4 hours		More than 4 hours	
	Wet	Dry	Wet	Dry	Wet	Dry	Wet	Dry
Yala	95.0	95.0	4.2	4.2	0.0	0.0	0.8	0.8
Wagai	100.0	95.0	0.0	5.0	0.0	0.0	0.0	0.0
Ukwala	86.9	65.6	11.5	29.5	1.6	4.9	0.0	0.0
Ugunja	91.9	77.8	8.1	22.2	0.0	0.0	0.0	0.0
Boro	70.2	64.5	28.4	32.6	1.4	2.1	0.0	11.1
Uranga	80.0	72.5	15.0	22.5	5.0	5.0	0.0	0.0
Bondo/Usigu	100.0	32.7	0.0	43.9	0.0	20.4	0.0	3.1
Rarienda	95.0	35.0	5.0	41.7	0.0	21.7	0.0	1.7
Madiany	75.4	39.3	18.0	23.0	6.6	32.8	0.0	4.9
District	88.3	64.2	10.0	25.0	1.6	9.6	0.1	2.4

Source: Field Data

PERCEIVED QUALITY OF DRINKING WATER AND WATER TREATMENT PRACTICES

The survey sought information from the respondents on perceived quality of drinking water taken from various sources and water treatment practices within the local community. The majority of the respondents (in excess of 70%) indicated that, with the exception of piped water supply systems, protected springs, wells and boreholes and roof catchments, most of the existing water supply sources are contaminated (polluted) and therefore not fit for human consumption. The remaining 30% of the respondents were non-committal and hence could not provide information on the perceived quality of drinking water from various sources.

Table 3.5: *Percentage distribution of methods used to store household water in the division*

Division	Plastic Buckets	Clay Pots	Metal Drums	Metal Buckets	Sufurias	Jerricans
Yala	15.1	46.2	0.8	0.8	0.0	37.0
Wagai	12.5	85.0	0.0	0.0	0.0	2.5
Ukwala	3.3	93.4	3.3	0.0	0.0	0.7
Ugunja	19.2	53.5	0.0	0.0	0.0	27.3
Boro	7.1	83.0	2.1	0.0	0.0	0.0
Uranga	12.5	87.5	0.0	0.0	0.0	0.0
Bondo/ Usigu	35.7	44.9	11.2	0.0	0.0	8.2
Rarieda	5.9	10.6	0.0	0.0	50.0	0.0
Madiany	3.3	82.0	1.6	1.6	1.6	9.8
District	12,7	65.1	2.1	0.3	5.7	10.3

Source: Field Data

Table 3.6: *Percentage distribution of use of household water in the division*

Division	Water Use	
	Livestock Watering	Cooking, Drinking and Washing
Yala	0.8	99.2
Wagai	2.5	97.5
Ukwala	1.6	98.4
Ugunja	1.0	99.0
Boro	0.7	99.3
Uranga	0.0	100.0
Bondo/Usigu	3.1	96.9
Rarieda	3.4	96.6
Madiany	3.3	96.7
District	1.8	98.2

Source: Field Data

Table 3.7, extracted from the Kenyan government's District Annual Report (1996), attempts a classification of water quality characteristics of the existing water supply facilities in Siaya District.

Despite the need for water treatment, especially of water from the unprotected sources, only 26.1% of the respondents in the district treat water by boiling, 9% by filtering, 4% by application of chemicals and 0.3% by traditional methods. The majority of respondents (61.5%) took no action to improve their drinking water quality. The practice of boiling water is most common in Uranga, Wagai and Bondo, where 47%, 40% and 34.1% respectively of the households boil drinking water.

Table 3.7: *Percentage water quality characteristics of the existing water supply facilities in the division*

Division	Clean and not Salty	Clean but Salty	Turbid and Salty	Dirty and Salty	Has Faecal coliform	Others
Yala	98.6	1.4	0.0	0.0	0.0	0.0
Wagai	63.4	11.4	9.1	0.0	0.0	6.0
Ukwala	73.9	16.9	4.0	0.5	1.7	3.0
Ugunja	81.7	0.8	12.9	0.0	3.8	0.8
Boro	64.6	5.9	13.4	1.5	3.7	9.9
Uranga	44.9	34.6	8.7	5.5	0.8	5.5
Bondo	19.3	24.8	42.2	0.0	2.8	11.1
Usigu	27.3	18.2	27.3	0.0	4.5	22.7
Rarieda	45.5	11.9	25.9	2.8	2.8	11.2
Madiany	42.2	3.4	42.2	0.0	7.8	4.3
Total	62.5	12.1	15.0	1.0	2.6	6.8

Note: - 62.5% of the existing water supply facilities in the district have good quality water (clean and not saline).
- Water from 15.0% of the existing facilities requires full treatment (turbid and saline).
- Yala division has the highest percentage (98.6%) of good quality water from its facilities.

Source: GOK, (1996)

The use of chemicals is common in Rarieda (20%) and Bondo (8%). Water treatment through filtering is mostly practised in Uranga (40%) and Bondo (16%). The households in the divisions that enjoy relatively better access to safe water like Ukwala and Ugunja do not normally treat water. This is because most of the households in these divisions have access to piped water supply systems and protected wells, boreholes and springs. Table 3.8 provides a summary of the water treatment practises in Siaya District.

RELIABILITY OF WATER SOURCES

During the field survey, the reliability of water sources was determined by asking the respondents to indicate the reliability of the various water sources they have access to. This is judged in terms of whether water is available all the time, most of the time or not available most of the time. This means that the water source dries up or that the quantity of water is not sufficient for all the consumers. In the latter case, the water source(s) may therefore be regarded as unreliable.

From the protected water sources in the district, the combined sources such as springs, wells, boreholes, piped supplies and roof catchments have a reliability of 62.1%. The most reliable protected source during the dry period are springs with a continuous reliability of 92.1%, followed by boreholes (77% reliability) and roof catchments (60% reliability). It should be noted that although springs are the most reliable, they are mainly concentrated in the northern parts of the district. The overall district-wide situation is that protected sources have a continuous reliability of 62.1% and a 37.4% frequent reliability during the dry period.

Table 3.8: *Percentage distribution of household water treatment practices in the division*

Division	Boiling	Filtering	Chemical	Traditional	No Action
Yala	24.4	5.9	0.0	0.0	69.7
Wagai	40.0	2.5	2.5	0.0	55.0
Ukwala	6.6	0.0	1.6	0.0	91.8
Ugunja	5.1	6.1	0.0	0.0	88.9
Boro	20.6	16.3	0.0	0.0	63.1
Uranga	47.5	40.0	2.5	0.0	10.0
Bondo/ Usigu	34.1	2.0	8.2	1.0	54.7
Rarieda	6.7	0.0	20.0	1.7	71.7
Madiany	29.5	1.6	0.0	0.0	68.9
District	26.1	8.7	3.9	0.3	62.5

Source: Field Data

Unprotected water sources are the most unreliable with district-wide reliability percentages of 56.2%, 31.6% and 12.3% continuous, frequent and unreliable categories respectively during the dry season. During the wet season, the reliability percentages are 73.8%, 19.5% and 6.8% continuous, frequent and unreliable categories respectively. Unreliability of unprotected water sources especially rivers, streams, dams and ground catchments are worse in the southern part and some parts of the central zones of the district. In these zones, such water sources usually dry up during the dry season, leaving the residents with very few options in terms of adequate and potable water supply sources. In such situations, water sources that survive drought conditions are usually dirty, turbid or salty.

SOCIO-CULTURAL FACTORS INFLUENCING THE CHOICE AND USE OF THE EXISTING WATER SUPPLY SOURCES

The concept and practice of the usual form of rural water supply development have revealed a number of limiting imperfections that characterise the management of water resources in most developing countries. In Siaya district, most of the rural water supply development endeavours are made on the basis of improving the supply system to be able to give the consumer a better, more reliable and up-to-date product. This approach has resulted in widespread apathy among the local population. Consequently, most of the existing water supplies, especially the piped water systems, are usually broken, misused or abused by the very people they were intended for. Furthermore, the purest, most reliable and most adequate means of water supplies are usually vandalised or abandoned in favour of impure, traditional supplies.

An alternative approach is to consider the "product" from the point of view of the consumer. This involves analysing what the consumers' needs and aspirations are in relation to water. How are they related to other needs and aspirations? Does the type of water supply and management system correspond with these perceptions?

Can the water supply and management system be developed and maintained by the beneficiaries themselves?

This philosophy can be described as the User Choice Approach and has been found to be the most appropriate for water supply development. This approach is based on the realisation that: First, in selecting alternative sources of water supply, individuals and communities provide their own frame of reference and choice criteria for assessing sets of trade-offs. Second, choices in the means of water supply recommended and/or delivered to communities by the water authorities are usually accepted with less commitment and zeal than the choices made by the communities.

Unfortunately Siaya District is currently grappling with the problems of inappropriate water supply development planning models which are oriented towards external conditions rather than local ones. Rural water supply development experiences have shown that water supply authorities usually start by identifying factors that have bearing on the improvement of delivery mechanisms. They mainly deal with the externals of water supply development processes and their measurable symptoms. Many of such theoretical development approaches that are based on measurable indices (e.g. the amount of water or number of facilities provided and percentage of area coverage) do not even address the issue of water supply management according to local socio-cultural conditions and capacities. This has resulted in the failure of well over 50% of rural water supply projects in Siaya District.

This study has also revealed that the individual or community choice of water source for domestic use is influenced by cognitive factors such as perceptions, awareness and judgement which are in turn dictated by long-term practical experience and knowledge. The figures given below provide an illustration of the decision making process in terms of individual frames of reference in the choice and use of different water sources within the community. Alternative water sources consist of roof catchment systems, boreholes and wells, springs, streams and rivers, ground catchments, dams, lake points and piped water supply systems. The users evaluated these water sources according to six main criteria of quality including taste. Other main criteria were quantity, technological feasibility, reliability, financial feasibility and accessibility (physical distance or effort required). Less critical but also influential were the personal relationships (bonding) i.e. the consideration of relationships with other people using the same source.

The resulting frame of reference provides a model for both kinds of criteria that could be employed in an individual water supply choice as well as the process by which these two kinds of criteria are linked together. From the results of the survey, it was noted that the perception of quality represents the major criterion by which alternative water sources are either eliminated or accepted for further consideration. However, in the dry season the water quality criterion becomes irrelevant as most decisions are mainly based on quantity, reliability and accessibility. Of those water sources that are perceived as acceptable in quality, financial feasibility and accessibility (physical distance or effort required) are usually taken into consideration before making the final decision on which particular source to use. As for those water sources meeting minimal requirements of perceived quality, financial feasibility and accessibility, more refined judgements are made concerning the technology used, the quality and the relationships with other people.

SOCIO-CULTURAL BELIEFS AND PRACTICES

Socio-cultural beliefs and practices are intrinsic to the way of life of any community. They include, *inter alia*, the norms and taboos that regulate behaviour and attitude towards the provision, choice and use of water supply facilities. Such norms and taboos may have both positive and negative implications on water supply development. During the survey, questions were therefore asked to establish the impact of such variables on water usage, choice of water sources, water ownership, water collection and water storage.

Traditionally, water was seen by the Luo community in Siaya District as sacred. It was a taboo to sprinkle water at someone while fighting. Since water was used as a solvent or an ingredient in the preparation of traditional medicines such as purgatives ("manyasi"), using water from someone else's house to mix the purgative was taboo. Water could also be used for cleansing purposes. For example, a man or woman who had committed adultery was required to take a bath before carrying his or her child. The period of mourning was believed to be a very special time and until all the rituals had been performed, a person was not allowed to bathe. As mourning was considered an act of cleansing, a person could not take part in the day-to-day activities before all the rituals were observed. Furthermore, a widow was not allowed to bathe around a water source before "tero" (the rite of wife inheritance) because this was believed to bring "chira" (curse).

The Luo community in Siaya District held beliefs that governed the vital activity of water collecting. Traditionally, it was the responsibility of women and young girls to fetch water. Water was mainly collected in water pots, and washing hands or bathing in these pots was not allowed. Widows were not allowed to collect water before the mandatory mourning period was over. It was also believed that breast-feeding women should not fetch water after sunset. Men were not allowed to carry water, especially on their heads.

Women were not allowed to pass through somebody else's homestead while carrying water. This was to safeguard against the possibility of tripping and falling because breaking a water pot is considered a bad omen. If a woman broke a water pot while carrying a baby on her back, cleansing had to be done through the use of a purgative ("manyasi"). A mother was not allowed to fetch water from the home of her married daughter. *Is this perhaps what is meant?)* A woman carrying water on her head was not supposed to meet someone directly. She was expected to move aside to let the other person pass before proceeding. It is worth noting that the Luo community attached a lot of importance to water as a commodity and hence also the importance of the water-pot.

Making water available for use is a major concern of the Luo community. The decision to construct a water point is mainly taken by men. That is, men decide where and how the facility is to be constructed. This was in line with the belief that men were the main decision-makers. Women were not allowed to own or construct water points. They could, however, assist by carrying away the silt from the pond, the dam, the borehole etc. or by preparing food for the men. Only barren women or women who had reached the menopause were allowed to participate in the construction of water facilities. Girls were not allowed to participate in the construction of water facilities in the village where they were born.

The responsibility of identifying the construction site and mobilising people to work on it is usually vested in a Council of Elders comprising of key opinion leaders, village heads, clan heads and other prominent old men. In case someone drowned in a well before the construction was completed, the well would be abandoned and an alternative site identified. It was believed that such a well would be a bad omen in the village. Among the Luo community dying in this manner would not be considered a natural phenomena but a sign that all is not well either with the victim, his family or the clan. In such cases the site would be cleansed, and the dead man would be buried next to the ill-fated and uncompleted well.

In the provision of water facilities as with every other activity, the roles of men and women were clearly defined. Men who have always been considered the chief decision-makers defined these gender roles. As expected, men were biased as they did not choose the duties that were recurrent but instead those that only needed to be done once. These duties included deciding where and how a water point was to be constructed. Men would than mobilise each other and construct the facility.

During the construction work, men would assist women in removing sand and other light duties. On completion of the facilities, women took over their management and maintenance. In essence, women therefore ended up with more work although by design, it appears men are overworked. The theory that men are breadwinners has therefore clouded the community's judgement concerning so many things in Siaya District. Once the construction work was done, women would take over by providing the household with water. Women's work in the daily fetching of water and even in many cases overseeing the general management and maintenance of water supply facilities undoubtedly gave them the lion's share of the work.

Some of the beliefs and practices discussed above still serve as useful regulations and guidelines for the general management of water resources within the community. However, most respondents felt that most of these beliefs are outdated and therefore need to be done away with. First, the belief that water fetching is strictly a woman's duty provides an example of such outdated beliefs. This is because the woman in most rural households in Siaya District is responsible for the family's welfare in the very important areas of health and nutrition. Because water collection is usually the first priority, healthcare and nutrition will suffer because the woman will put all her energy in water collection. If men could be involved in water collection using locally available transportation methods such as handcarts, women would have more time available for other useful purposes such as primary health care and ensuring that the nutritional needs of the family are met. Second, the tradition of barring widows who have not undergone the ritual of wife inheritance from using communal water sources was considered both outdated and unrealistic. The AIDS scourge has forced the Luo community to think twice about wife inheritance. This means therefore that there are more widows in the community who have not been inherited. If this tradition were preserved, more and more people in the community would be barred from using water. Such outdated traditions as well as others not mentioned here should be identified and either made relevant to current local conditions and circumstances or completely done away with.

COMMUNITY PARTICIPATION IN RURAL WATER SUPPLY DEVELOPMENT

Recently it has been agreed upon in general that rural water supply agencies should pursue the course of "African socialism". The participation of all people in the task of water supply development as well as in the achievement of their collective needs was sought as the only means to reach these goals. Unfortunately, these goals were poorly articulated. With the exception of rural water supply projects run by the Kenya-Finland Cooperation Agency (Kefinco), the planning and administration of rural water supply development projects in Siaya district has been highly centralised. Most of the rural water supply projects have therefore failed because up-down strategies have been adopted for improving water supplies. Under these strategies, teams of experts linked to central authorities come into each village, decide what is needed, plan and construct the necessary facilities, acquire and install the required equipment, and make arrangements for centrally supported, regular maintenance. Access to the improved water supply is free or almost free to everyone in the village, along the philosophy that no charge should be collected for an essential, basic need such as water. Villagers are also not expected to contribute significant resources to the effort in any other way, therefore virtually all of the costs are covered by funds from outside the village. Practically the only role for the community is to receive and use the finished product.

The hierarchical structure of the various agencies involved in rural water supply development has resulted in most decisions having been made at the headquarters and passed down to the field staff. There is too little horizontal integration - water supply projects for villages are prepared, priorities worked out, budgets allocated, and personnel deployed by the sector agencies. As a result, much of the rural water supply development planning has been based on inadequate knowledge of local environmental conditions or constraints. The outcome of such water projects has already fallen far short of expectations. In fact, a significant proportion of rural water supply improvement projects in Siaya District, especially piped water supply facilities in the southern parts of the district covering Madiany, Rarieda, Bondo and Usigu divisions, are not longer operational. Large investments have therefore brought few lasting results. Most of the intended beneficiaries (in excess of 80%) interviewed indicated that they had never been involved in any stage of a water supply development planning process. However, some cases were noted where the community has been making contributions (in terms of money, labour and materials) towards rural water supply development especially during implementation, maintenance and operation (table 3.9). In most cases, the ideas, needs, values, aspirations and aversions of the local community and the existing traditional knowledge and water resources are integrated. Unfortunately, the current resource anagement systems are usually not integrated in the rural water supply development process. The emergence of such a scenario could be attributed to attitudinal variables on the part of the community - to weak or non-existing institutional frameworks through which the local peoples could be empowered. Hence people were made to participate effectively in matters concerning their welfare - in the lack of understanding of the unmet needs of the society and how such needs could be integrated in the water planning process.

Getting local people involved in the planning of community water supply development activities should instil a sense of ownership thereby leading to greater commitment and support. Considerable evidence has been amassed that shows that local participation in water supply development planning is crucial to acceptance and sustained support for rural water supply development.

A participatory water development strategy has to be complimented by local people's capacities through mobilising their energies. Such local capacity is indigenous local knowledge. This is more likely to come about when there is genuine local participation in the water planning process. To this end, water planners must address the *unmet needs* of the society, which are the most powerful motivating forces behind genuine and effective community participation in water supply development endeavours. Such needs also include the substantive, procedural and psychological needs.

Table 3.9: *Nature and percentage community contribution during, implementation maintenance and operation*

Division	Not Applicabl	Money 1	Labour 2	Materials 3	All the three 1,2,3	Others
Yala	0.0	4.2	90.5	4.2	1.4	0.0
Wagai	8.6	8.6	49.1	10.3	5.7	17.7
Ukwala	1.0	33.6	45.5	0.7	2.0	17.2
Ugunja	11.3	1.3	62.1	2.1	2.5	20.8
Boro	11.4	11.1	53.5	7.0	14.0	3.8
Uranga	6.3	3.1	33.1	0.0	41.7	15.7
Bondo	9.2	18.3	69.7	0.9	0.9	0.9
Usigu	0.0	40.9	36.4	3.0	1.0	0.0
Rarieda	4.2	7.0	58.7	2.1	16.8	11.2
Madiany	13.8	14.7	50.0	0.9	20.7	0.0
District	7.0	14.7	53.4	3.2	10.0	11.7

Note: - Labour contribution by the community has the highest percentage (53.4%).
- Yala division has the highest percentage of labour contribution (90.5%).

Source: Field Data.

The first and most easily recognised *substantive* is the need for communities to know about "what" a proposed idea or action is, that is, the "substance" of what is being suggested. The first need to interest and affect a community is to have information about the proposed water project. Water planners are usually fairly competent at meeting substantive needs. Unfortunately, they tend to be unaware of the other needs. This is usually where dissemination of information about the proposed water project to the community stops, leaving two further unmet needs. Those unmet needs are where the failure (or trouble) of the water project starts.

The second need that might not be met is the need to be part of the process of considering and discussion – the "how" – the substance of the proposal – the "what". The needs to determine "how" decisions are taken and about "What" are referred to, as the *procedural needs*. The interested and affected parties have the need to be part of a process that considers whether a water project is needed or wanted. If it is, then where it should be and what are the requirements and who should take part in construction and maintenance? There is a great deal of energy and power, both positive and negative, locked up in procedural needs. When the local people have had the opportunity to have their say and contribute their ideas, when the local people have been part of a decision, there is a far greater chance that it will liberate their sense of ownership of that decision. This is based on the philosophy that something discussed with you becomes part of you. If you feel you "own" something you are far more likely to nurture and care for it. Meeting procedural needs can therefore help to engage and focus this personal energy and commitment to rural water supply development in Siaya district. However, there is also negative energy and power locked up in procedural needs that are not met. It should be remembered that one of the greatest insights in the field of human motivation is that satisfied needs do not motivate people to participate effectively in any given project. It is only the unsatisfied need that motivates. When you, as a planner, do not meet someone's procedural need you risk the danger of unleashing a large store of energy that can motivate and drive behaviour. This behaviour is usually in opposition to your proposal and hence you would view it as negative. Most of the rural water supply projects in Siaya District were conceptualised, planned designed and implemented without consulting the intended beneficiaries thereby neglecting their procedural needs. Such projects are then vandalised, under-utilised and abandoned (table 3.10). This is Probably the energy released by that unmet procedural need is shaping the behaviour of the local population in a way that opposes the proposed or existing water project.

Table 3.10: *Percentage distribution of water supply facility status in they division*

Division	Completed and Functioning	Completed But Not Functioning	Under Construction	Abandoned/ Stalled	Broken Down
Yala	88.5	5.8	1.9	0.0	3.8
Wagai	60.7	8.1	20.2	0.0	11.0
Ukwala	90.1	5.3	1.4	0.3	2.9
Ugunja	88.8	3.2	1.8	2.2	4.0
Boro	80.0	1.7	10.0	2.4	5.9
Uranga	65.0	9.4	3.4	6.8	15.4
Bondo	44.8	10.6	16.7	15.3	12.6
Usigu	38.1	23.8	33.3	0.0	4.8
Rarieda	41.5	10.0	17.0	15.4	16.1
Madiany	49.0	16.0	1.1	14.0	19.9
District	78.2	6.1	8.0	2.5	5.2

Source: Field Data

The third need is the deep and profound need of every human being to be valued and regarded as worthy consideration. This third need is called the *psychological need.* Psychological needs are strongly linked to procedural needs because of their amplifying effect. If the water planner do not make the effort to involve people in discussions about the proposed action, he sends them a message that he does not value them sufficiently to bother to talk to them. The people may also interpret this to mean that the planner thinks that they are not capable of contributing to problem solving through consensus building. The feelings of the people is that the planner is proposing to do something that will affect their lives and yet they are not even given the opportunity to discuss it.

ABILITY AND WILLINGNESS TO PAY FOR IMPROVED WATER SUPPLY SERVICES

According to the results of the survey, Siaya District is generally a poverty-stricken area characterised by households having an income of less than Kshs. 1000 per month. Tables 3.11 and 3.12 give the break down of the monthly and annual average household income distribution by division. The results showed that about 85.7% of the households interviewed fall within the mean income bracket ranging between Kshs. 0 to 1999, 12.8% in the income bracket of between Kshs. 2,000 to 7,999 and 1.4% in the over kshs. 8,000 income category. The income groupings were computed using the Central bureau of Statistics classification.

The field results also illustrate that the majority of households in the district fall within the income bracket of Kshs. O to 1999. Ugunja division has majority of its household members of (86%) in this cadre with relative frequency of 99%. Madiany division has the least of its household members earning between Kshs. o to 1999. Most households in Madiany have higher incomes between Kshs.

Table 3.11: *Monthly and annual average household income by division*

		Income by Source		
Division	Mean Income	Farm/ business	Wage	Total
Wagai	Monthly	1034.50	340.00	1374.50
	Annual	6642.25	3085.00	8727.25
Ukwala	Monthly	627.00	462.20	1089.20
	Annual	7771.60	5375.10	13146.70
Boro	Monthly	621.00	372.60	993.60
	Annual	5834.75	3359.20	9193.95
Bondo/Usigu	Monthly	798.20	279.70	1077.90
	Annual	6182.30	4092.00	10274.30
Madiany	Monthly	908.10	1571.20	2425.30
	Annual	18132.30	8318.60	26450.80
Ugunja	Monthly	133.00	3.50	136.50
	Annual	1756.45	42.00	1798.45
Rarieda	Monthly	508.85	5000.80	1009.65
	Annual	6437.80	5400.00	11837.80
Yala	Monthly	469.30	373.20	842.50
	Annual	4666.60	3020.00	7686.60
Uranga	Monthly	315.40	0.00	315.40
	Annual	24448.60	0.00	2448.60
District	Monthly	572.90	402.50	975.40
	Annual	6196.30	3490.30	9686.60

Source: Field Data

2000 to 7,999 with 25.4% of the households in this category of income. Ugunja division has the least proportion of its households with monthly income between Kshs. 200 to 7999 with only 1% of the households earning this income. The other divisions following Madiany with households earning between Kshs. 2000 to 7999 are Wagai, Bondo, Rarieda, Yala, Ukwala, Boro, Uranga and Ugunja (table 3.12).

As noted earlier, Siaya District is generally a poverty stricken area characterised by households with an income of less than Kshs. 1000 per month. taking 5% of the monthly income as the Rule of Thumb for determining the affordability level of the community, it can be argued that the resultant figure of KShs. 50 per month may be affordable for most of the households though it may be "too expensive" for households with very low monthly income.

Despite the generally low monthly income in Siaya District, a significant proportion of respondents (in excess of 70%) expressed their willingness to pay for improved water supply services. However, substantial differences appeared concerning the amount of money people are willing to pay for improved water supply services. Respondents from the northern parts of the district comprising of Ukwala and Ugunja divisions were willing to pay significantly greater average amounts per households per month (Kshs. 50. To 100) than their counterparts in the central and southern parts which comprise of Uranga, Boro, Wagai, Yala, Bondo, Rarieda and Madiany divisions.

It was noted during the survey that those respondents who ae willing to pay for improved services have either been using such services or have decided to install them individually in their homes or as a joint effort by the

Table 3.12: *Percentage distribution of household income by division*

Division	Amount in Kshs. 0-1999	Amount in Kshs 2000-7999	Over Kishs 8000	Number of cases
Wagai	80.0	20.0	-	21
Ukwala	86.9	11.5	1.6	32
Boro	85.1	11.3	3.5	74
Bondo/Usigu	83.0	16.0	1.0	52
Madiany	69.5	25.4	5.1	32
Ugunja	99.0	1.0	-	52
Rarieda	85.2	13.1	1.6	32
Yala	87.7	12.3	-	64
Uranga	95.0	5.0	-	21
District	85.7	12.8	1.4	380

Source: Field Data

community. Such improved water supply services include, *inter alia*, improved borehole and spring supplies and pied water supply systems, which were mainly initiated by Kefinco and Care-Kenya among other stakeholders. Those respondents with minimal exposure to improved services were reluctant to commit themselves that they are willing to pay for such services. Some of hem do not even see the danger of using traditional sources of water. In fact, in some divisions such as Rarieda and Madiany most of piped water supply facilities initiated by the government have either been abandoned, or vandalised by the very people they were intended to serve. In such cases the local people resorted back to the use of traditional sources of waters supply irrespective of their poor quality. Some people even argued that it is the responsibility of the government to provide such services. Though such negative attitudes towards rural water supply development are not very common in Siaya district, it is believed that such information could be useful to the water resource manager when choosing among water planning alternatives. This kind of investigation was aimed at discovering or unearthing what individuals 'see' in their environment, how they feel about it, what it means to them, and how they would be disposed to act towards it. Such information should form an integral component of the water planning process.

INSTITUTIONAL ARRANGEMENTS

One of the least touched upon, but possibly one of the most fundamental research needs in resource management is the analysis of how institutional arrangements are formed, and how they evolve in response to changing needs and the existence of internal and external stress. Structure and operational guidelines by which resource management institutions are formed and evolve clearly affect the implementation of resource policy, both as to the range of choice adopted and the decision attitudes of the personnel involved (O'Riordan, 1971).

For the purposes of this chapter, institutions and institutional arrangements refer to a definable system of public decisions making, which includes specific organisational entities and jurisdictions. In addition to being concerned with component organisational entities, the term institutions suggest special attention to the configuration of relationships which are: (i) established by law between individuals and government; (ii) involved in economic transactions among individuals and groups; (iii) developed to articulate legal, financial and administrative relations among public agencies; and (iv) motivated by socio-psychological stimuli among groups and individuals. Such relationships describe the institutional system for decision making. Thus, institutional arrangement investigations of this nature focus on the linkages, which bring authorities or agencies and action centres together into a public decision making system, which is responsive to the socio-cultiral environment within which it must operate. Simply stated, institutional arrangements may be defined as a cluster of customs, laws, or ways of behaving and organising behaviour around problems of life in society. They may also be viewed as forms of government, agencies, legislation, and other socio-cultural guides established to influence human behaviour. The concept of institutional arrangements can therefore be used as an umbrella term to incorporate behavioural, managerial and other socio-cultural variables that may have an influence on rural water supply development. However, it should be noted that institutional arrangements represent only one of many variables that constrain or influence water resource management process.

The major stakeholders in rural water supply development include, *inter alia,* Kenaya government, Lake Basin Development Authority (LBDA), Kenya-Finland Co-operation Agency (Kefinco) and Care-Kenya. With the exception of water projects initiated by Kefinco, most of the other projects run by the other agencies have been experiencing a lot problems which could mainly be attributed to the operations of the institutions involved and technology in use. Until recently, the administration of rural water supply projects in Siaya District has been highly centralised. This has resulted in several problems in the design and implementation of various specific projects. First, the hierarchical structure of the various agencies (institutions) involved in rural water supply development has resulted in most decisions having been made in the respective headquarters and passed down to the field staff. As a result, much of the rural water supply development planning has been based on inadequate knowledge of local environmental constraints. Second, different rural water supply projects have been the responsibility of different agencies. Given the generally inadequate inter-institutional co-ordination, many water supply services have been duplicated while others have been completely omitted. Third, this study revealed that while there has been general agreement that the agencies should fully involve the intended beneficiaries in all stages of the water supply project cycle, the specific means or strategies for achieving this goal have often been poorly articulated. It should be noted that the opportunity to participate in rural water supply development depends on the strength of the institutions operating at the local level and their modes of operation. For instance, a public meeting (a *baraza*) at the location or even sub-location level can be used as a platform at which district level officers could elicit the views of the local people. Alternatively, experience has shown that it can also be used solely to issue directives top-down. If it does, the latter, a situation will arise where water projects are centrally planned and designed by the agencies without thorough knowledge of the local economic, socio-cultural and physical conditions and without the local people's involvement.

As a result of inadequacy in institutional structures for managing water resources, a number of factors were identified as causing an imperfect response to rural water supply problems. These include (i) a diffused public interest, (ii) differing views about priorities, (iii) inadequate legislation and enforcement, (iv) special interest politics, (v) fragmentation of responsibilities, (vi) organisational jealousies, and (vii) misunderstanding of man-environment relationships.

Institutions charged with the responsibility of rural water supply development in Siaya district are characterised by weak communication and co-ordination mechanisms between (i) various agencies, (ii) field personnel and researchers involved in water planning, (iii) the central and the local-level bureaucracy, (iv) the bureaucracy and the local people, and (v) national administrators and donors. The existing administrative structures and procedures are inadequate to handle the conceptualisation, planning, design and implementation of rural water supply development projects. First, there is too little horizontal integration – that is, water projects for villages are prepared, priorities worked out, budgets allocated, and personnel deployed by the institutions (agencies) concerned with rural water supply development. Field staff are usually responsible to their parent agencies and rarely acted as members of a unified development team. Second, the vertical set-up of these institutional bureaucracies with too little lower echelon input. Decisions usually take an inordinately long time to reach the field personnel and often overlooked local resource endowments and environmental factors. Consequently, local community initiatives and co-operation are often strangled. These problems are further aggravated by the fact that the operations of such institutions are constrained by inadequacy of financial, technical and manpower resources. The limited trained local skilled labour creates further difficulties with the management of water resources. Weak or lack of institutional structures and procedures therefore form major constraints to sustainable rural water supply development in Siaya District.

To this end, there is need for proper and effective institutional structures and procedures to be put in place so as to promote meaningful consultation with rural residents, careful identification of local needs, local resources and constraints, intelligent application of appropriate technical and institutional criteria. These have to be supported by a feedback mechanism to assess progress and to remove obstacles in the way of its realisation. However, experience from Siaya District indicated that the successful application of such institutional structures and procedures is frequently blocked by bureaucratic inertia and political pressures that usually result in adopting or abandoning certain water supply projects, irrespective of their intrinsic merit or demerit.

Despite regular District Development Committee (DDCs) meetings to review water supply projects and recommend alternative measures for rural development in their respective regions, indicators of water supply development, either pointed towards deterioration or stagnation. It should be noted that most of the failures in rural water supply development in Siaya district are rooted in administrative and political problems (the district being located in an opposition stronghold area) as well as economic and other socio-cultural problems. There is therefore need to reform administrative and political systems so that they respond more effectively to the needs and demands of clientele groups.

CONCLUSIONS

During the past three and half decades, Siaya District has seen an unprecedented surge in rural water supply development projects/programmes aimed at providing solutions to inadequacies of potable water supply, water resource management and other water-related problems. Despite these efforts, inadequacies of potable water supply and mismanagement of water resources have continued unabated in most parts of the district. Consequently, water supply problems are becoming increasingly severe and widespread as the quality of life deteriorates and the livelihood of thousands of people is threatened. Ironically, according to information obtained during field survey, most of the rural water supply projects (in excess of 50%). This especially so for activities initiated by the government that have either been abandoned, vandalised or only function intermittently and have therefore failed to meet the expectations of the intended beneficiaries, government, donors and other interested and affected parties. Such project failures could mainly be attributed to inadequacies or weaknesses of rural water supply development approaches adopted. These include, among other things, adoption of top-down approach which is supply-driven rather then demand-driven; lack of meaningful community involvement in the entire project cycle; use of inappropriate technology; and, lack of understanding of the local environmental conditions.

Most rural water supply development projects/programmes have focused their activities on delivery mechanisms and quick technical solutions. Fundamentally lacking in these approaches to incorporate local knowledge, experiences, technologies, aspirations, aversions, skills, wisdom, institutions and governance systems and other socio-cultural elements. The inability of these water supply projects to include analyses of social and cultural phenomena, which obviously have profound influence on the relationship between people and their environment, has contributed to their failures. It is increasingly apparent that it takes more than efficient delivery mechanisms and quick technical solutions to achieve sustainability in water supply development. Furthermore, by de-emphasising the links and relationships between people, their socio-cultural characteristics and water resources, previous and ongoing waters supply development projects have, *inter alia*, unwittingly contributed to the currently prevalent insensitivity and apathy of communities towards sustainable development of rural water supply development in Siaya District. With the downplaying of cultural values, norms, aspirations, indigenous knowledge systems, skills and technologies, the people's ability and will to manage water resources sustainable for supporting their livelihood has not been fully realised. It is therefore, recommended that the local communities, women and youth groups be fully involved in water supply development and management at all levels.

The existing rural water supply development institutions and technology in use are grossly inadequate in terms of their feasibility, flexibility and acceptability. Such institutions are characterised by outdated water supply design standards, which do not involve the local community adequately and usually fail to take account of cultural practices and customs as they relate to water supply development. This therefore calls for thorough review of the mandate and roles of the various institutions involved in the development and management of water resources with a view to streamlining their roles, responsibilities and operations as well as strengthening their capacities. Such review should include, delineation of institutional responsibilities and clear statement of operational terms of reference so as to avoid duplication of efforts and conflicting interests.

One of the main priorities in the broad field of rural water supply development should be capacity building and institutional strengthening. Even if excellent water delivery mechanisms exist and economic incentives are in place, it only with the presence and assistance of appropriate institutions that any water supply project can succeed. This includes elements such as:

(i) The presence of individuals capable of designing appropriate water supply projects, policies and regulations, and of technical experts who can mitigate environmental problems and monitor and evaluate project performance;
(ii) The existence of flexible national, regional and local environmental authorities within government structures with legal power to enforce environmental laws and make decisions at different levels;
(iii) Reliable laboratories, equipment, infrastructure and economic resources for field monitoring, assessment and research. For instance, this will enhance institutional capacities to assess and determine water quality and groundwater potential throughout the district. This will avoid the current practice of wasting scarce resources in siting and developing boreholes in areas with very low groundwater potential. Strengthen the various water testing laboratories and field survey capacity through the provision of adequately trained personnel and necessary laboratory and field equipment. These in turn would strengthen laboratory and field operations, especially water quality monitoring and hydrogeological surveys;
(iv) The presence of social institutional framework (at the grass root level) through which the cultural values, norms, aspirations, aversions and indigenous knowledge systems could be identified, analysed and integrated into the water planning process. Such institutions will play both a facilitating and empowerment role as well as creating an enabling environment for the intended beneficiaries to air their views, values and fears as they relate to water supply development; and,
(v) The existence of training and environmental awareness creation programmes within the institutional structure. This will assist in developing the local human resources and therefore empowering them to take full responsibility of he management, maintenance and administration of water supply projects. Such programmes will also in-still a strong sense of awareness of the dangers of using unprotected traditional water sources.

Experience from Siaya District has shown that top-down and narrow sectoral approach to water resource development and management cannot work. Such approach is characterised by: (i) a sectoral focus on a single use and single user; (ii) a prescriptive approach, usually bringing technology and knowledge which is not culturally or ecologically appropriate; (iii) a disregard for other uses and users of the target water resource or the area it occupies; and, (iv) a neglect or active exclusion of local people and communities from any part in the project cycle. Consequently, the local people, who for the most part are not effectively consulted nor involved in any part of the project cycle, usually resent, ignore, under use, abuse or occasionally destroy the very water supply facilities and resources which are intended to serve them. This approach has proved ineffective in solving the multifarious water supply development problems facing Siaya District. Such approach pays no attention to local people's knowledge or socio-cultural values, which should form an integral component of water supply development.

To be sure, an integrated water planning approach is highly recommended if meaningful and sustainable rural water supply development is to be realised this approach calls for informed decision-making. These are coupled with environmental accountability which is achieved using an open, participatory approach of public and specialist input during all stages of the water planning process, with due consideration of alternatives and a genuine attempt to mitigate negative impacts and enhance positive aspects. Integrated water planning is also based on: (i) systematic consideration of the various components of water which include surface and ground water, quantity and quality; (ii) recognition of the fact that while water is a system, it is also a component which interacts with other systems; and (iii) due consideration of the interrelationships between water supply development and socio-cultural variables.

BIBLIOGRAPHY

Baker, D. (1979) The Supply of Water to Rural areas of Kenya for Domestic purposes. In: Castelino, J. B. and Khamala, C. P. M (eds.), *The Role of Water Resources in Development*. Proceedings of the 13th annual symposium of the East African Academy; Printing and Packaging Corporation Ltd., Nairobi, pp. 192-199.

Ball, C. and Ball, M. (1991) *Water Supplies for Rural Communities*. Intermediate Technology Publications, London, p. 56.

Biswas, A. K. (1974) Water, In: Sargent 11, F. Ed.) (1974). *Human Ecology*. North-Holland Publishing Company, New York, pp. 207-232.

Biswas, A. K. (1993) Management of International Waters: Problems and Perspective. In: Biswas, A. K. (ed.) (1993). *International Journal of Water Resources Development. Environment and Water Development: Some Critical Issues*. Vol. 9, No. 2, 1993. Carfax Publishing Company, United Kingdom, pp. 197-181.

Briscore, J. and de Ferranti, D. (1988) *Water for rural Communities: Helping people Help Themselves*. The World Bank, Washington, D.C., 32 p.

Carruthers, I. D. (1972). *Impact and Economics of community Water Supply: A Study of rural Water Investment in Kenya*. London, Wye College Publishers, p. 165p.

Churchill, A. A.; de Ferranti, D,; Roche, R,; Tager, C,; Walters, A. A. and Yazer, A. (1987) *Rural Water Supply and Sanitation: time for a Change*. World Bank Discussion papers. The World Bank, Washington, D.C., p.111.

Dangerfield, B. J. (ed.) (1983). *Water Supply and Sanitation in Developing Countries*. The Institution of Water Engineers and Scientists, London, pp. 1-5, p. 239, pp. 207-329.

Donaldson, D. (1977) Progress in the Rural Water Programmes of Latin America. In: Feachem, R.; McGarry, M. and Mara, D. (eds.) (1977). *Water, Wastes and health in Hot Climates*. Jhon Wiley and Sons, New York, pp. 213-227.

Donaldson, D. (1988) Community Participation in urban Water Supply and Sanitation Systems. In: Aquatech 91988). *Developing World Water*. Proceedings of the 12th International Exhibition and Conference on Water Technology, Amsterdam, Netherlands, pp. 86-89.

Dube, S. C. (1988) *Modernisation and Development: The Search for Alternative Paradigms*, Studies on Socio-cultural development alternatives in a changing world. London: Zed Banks Ltd., pp. 1-114.

Feachem, R,; Burns, E,; Cairncross, S,; Cronin, A,; Cross, P.; Curtis, D,; Khan, M. K. Lamb, D,; and Southall, H. (eds.) (1978) *Water, Health and Development: An Interdisciplinary Evaluation*. Tri-med Books Ltd., London, p. 267.

Gans, H. J. (1974) Planning for People, not buildings. In: Sargent 11, F. (ed.) (1974). *Human Ecology*. North-Halland Publishing Company, New York, pp. 305-429.

Golubev, G. N. (1993) Sustainable Water Development: Implications for the Future. In: Biswas, A. K. (ed.), *International Journal of Water Resources Development, Environment and Water Development: Some Critical Issues*. Vol. 9, No. 2.Carfax Publishing Company, United Kingdom, pp.127-152.

Habitat (1982) *Appropriate Infrastructure Services, Standards and Technologies*. Report of the Ad Hoc Expert Group Meeting on Appropriate Infrastructure Services, Standards and Technologies for Upgrading Slums and Squatter Areas and Rural Settlements, Nairobi, Kenya, p. 68.

Habitat (1986) *Delivery of Basic Infrastructure to low-income settlements: Issues and Options*. Habitat Nairobi, Kenya, p. 48.

Habitat (1989) *The Conservation of Drinking-Water Supplies: Techniques for low-Income Settlements*. Habitat, Nairobi, Kenya, p. 48.

Jacobsen, B.; Ascroft, J,; and, Padfied, H. (1971) *Strategies for Improving Rural Welfare: The Case for Rural Water in Kenya*. Unpublished Occasional paper No. 4. Institute of Development Studies (IDS), University of Nairobi.

Japan International Co-operation Agency, (1991) *The Study on The Kenya National Water Master Plan: Report No. 1*. Unpublished Report, Ministry of Water Development, Nairobi, pp. 180-187.

Kenya, Republic of, (1996) *Siaya District Annual Report*, Unpublished Report, p. 288.

Kenya, Republic of (1996) *Water and Sanitation Baseline Survey Report for Siaya District*, Unpublished Report.

Kenya, Republic of (1993) *Siaya District Development Plan*, 1993-1996, Government Printer, Nairobi; pp. 1-39.

Kenya, Republic of (1988). *Siaya District Rural Domestic Water Resources Assessment Report,* December 1988, Unpublished Report, p. 288..

Kenya, Republic of, (1985). *National Demographic Survey*. Government printer, Nairobi, pp. 1-7.

Kenya, Republic of (1980) *National Master Water Plan; Summary and National Plan: Volume 1*. Nairobi, Government Printer, pp. 9-14.

Kenya, Republic of (1989) *Siaya District Development Plan*, 1989-93. Government Printer, Nairobi.

Kenya, Republic of (1994). *Siaya District Development Plan, 1994-1996.* Government Printer, Nairobi.

Kenya, Republic of (1989) *National Development Plan, 1989-1993.* Government Printer, Nairobi.

Kenya, Republic of (1984) *National Development Plan, 1984-1988.* Government Printer, Nairobi.

Kenya, Republic of (1979) *National Development Plan, 1979-1983.* Government Printer, Nairobi.

Kenya, Republic of (1974) *National Development Plan, 1974-1978.* Government Printer, Nairobi.

Kenya, Republic of (1970) *National Development Plan, 1970-1974.* Government Printer Nairobi.

Kenya, Republic of (1966) *National Development Plan, 1966-1970.* Government Printer, Nairobi.

Khair, T. and Islam, A. S. M. (1990) Rural Water Supply Development in Bangladesh. In: Asian Development Bank (ADB) and United Nations Development Programme (UNDP) (1990) *Women and Water: Domestic Shallow Well Water Supplies: The Family Handpump Scenario.* Proceedings of Regional Seminar, Manila, pp. 71-93.

Krhoda, G. O. (1988). Water Supply in Kenya Today and to the Year 2000 AD. In: Ominde, S. H. (ed.), *Kenya's Population Growth and Development to the Year 2000 AD.* Heinemann, Nairobi, pp. 87-99.

MacGarry, M. G. (1977) Institutional Development for Sanitation and Water Supply. In: Feachem, R,; McGarry,M. and Mara, D. (eds.) 91977). Water, Wastes and Health in Hot Climates. John Wiley and Sons, New York, 195-211. McHale, M. C. (1982) An Opportunity in Interdependence: The Fulfilment of Basic Needs. In: Mayor, F. (ed.) (1982). *Scientific Research and Social goals: Towards a New Development Model.* Oxford: Pergamon Press, pp. 233-236.

Mclntosh, A. C. (1990) The Family Handpump. In: Asian Development Bank (ADP) and United Nations Development Programme (UNDP) (1990). *Women and Water: Domestic Shallow Well Water Supplies: The Family Handpump Scenario.* Proceedings of Regional Seminar, Manila, pp. 35-52.

Mitchell, B. (1979) *Geography and Resource Analysis.* Longman Group Limited, London, 1-3; pp. 118-143

Nedge, M. (1993) *Water Resources Development, Uses and Potentials.* Unpublished Seminar Paper Presented to the Nyanza Leaders Conference held in Kisumu, Kenya, p. 22.

Nyaoro, W. (1990) *The Distribution and Variation in the use of Primary Schools: A Case Study of Bondo Division, Siaya District, Kenya.* Unpublished B.Sc. Dissertation, University of Nairobi.

Nyaoro, W. (1992). *Water Supply and Consumption in Eldoret.* Unpublished M. Phil. Thesis, Moi University.

Ongwenyi, G. (1993). *Nyanza Province: Towards Water Development, Management and Supply.* Unpublished Seminar Paper Presented to the Nyanza Leaders Conference held at Kisumu, Kenya, pp. 1-9.

O'Riordan, T. (1971) *Perspectives on Resource Management.* London: Pion Publishers Ltd., pp. 1-9.

Overman, M. (1986) Water: *Solutions to a Problem of Supply and Demand.* The Open University Press, p. 15.
Owen, L. D. and Goldsmith, P. L. (eds.) (1972) *Statistical Methods in Research and Production.* Longman Group Ltd., London, pp. 121-229.
Panikkar, R. (1992) Philosophical Investigation of Sustainable Development: Fundamental Issues. In: UNEP (1992): *Living with the Earth*, Nairobi: UNEP Publications Ltd., pp. 28-30.
Poulton, R. and Harris, M. (1988) *Putting People First: voluntary Organisations and Third World Organisations*, Development Studies Series. London: McMillan Publishers Ltd., pp. 11-85.
Scotney, N. (1976). *Evaluation of the Rural Water Supply programme: Report on Some Relevant Social Factors Based on a Survey of Three Rural Water Supply Schemes in Embu, Kitui and Kakamega Districts.* Nairobi, Government Printer, pp. 1-23.
Sen, G. and Grown, C. (1988) *Development, Crises and Alternative Visions: Third World Women's Perspectives.* London: Earthscan Publications Ltd., 14-49, pp. 74-95.
Simpson-Herbert, M. (1990) Women and Water. In; Asian Development Bank (ADB) and United Nations Development Programme (UNDP) *Women and Water; domestic Shallow Well Water Supplies: the Family Handpump Scenario*: Proceedings of Regional Seminar, Manila, pp. 55-69.
Soedjatmoko, K. C. (1994) *Transforming Humanity: The Visionary Writings of Soedjatmoko.* Connecticut: Kumaran Press, 1-75.
Thanh, N. C. and Biswas, A. K. (1990) *Environmentally Sound water Management.* New York: Oxford University Press.
United nations (1977) *Report of the United Nations Water Conference Held at Mar del Plata.* Argentian, p. 181.
United Nations (1980) *Report of the World Conference of the United Nations Decade for Women: Equality, Development and Peace. Copenhagen. United Nations Publications*, pp. 1-16.
UNESCO, (1986) *Strategies for Endogenous Development.* New Delhi: Oxford and IBH Publishing Co., pp. 1-73.
UNICEF (1993) *Water Supply: focus on Homa Bay.* Unpublished Seminar Paper. Unicef, Nairobi, pp. 1-3.
White, A.U. Bradley, D. J. and White, G. F. (1972) *Drawers of Water: Domestic Water Use in East Africa.* University of Chicago Press, Chicago, pp. 7–151.
White, A. U. (1977) Patterns of Domestic Water Use in Low-Income Communities. In: Feachem, R,: McGarry. M. and Mara, D. (eds.) (1977*). Water, Wastesand Health in Hot Climates.* John Wiley and sons, New York, pp. 96-111.
White, A.U. and Burton, I. (1977) Water Supply and Community Choice. In: Feachem, R,: McGarry. M. and Mara, D. (eds.) (1977*). Water; Wastes and Heath in Hot Climates*: John Wiley and Sons, New York, pp. 113-129.
Whiting, M. (1979). A Catalyst for the Implementation of Self-Help Water Projects. In: Castelino, J. B. and Khamala, C. P. M. (eds.), *The Role of Water resources in Development.* Proceedings of the 13th annual symposium of the East African academy; Printing and Packaging Corporation Ltd., Nairobi, pp. 273-277.
World Health Organisation (WHO) 1981*) Drinking-water and Sanitation, 1981-1990: A Way to Health,* Geneva, pp. 56.

World Health Organisation (WHO)(1983) Progress in the attainment of the goals of the International Drinking Water Supply and Sanitation Decade–Report of the Secretary General. In: United Nations Centre for Human Settlements (Habitat) (1986) *Delivery of Basic Infrastructure to Low-income Settlements: Issues and Options* pp. 4-9.

World Health Organisation (WHO) (1985) *The International Drinking Water Supply and Sanitation Decade – World Health Statistics 1985*. Geneva, pp. 1-7.

4 Technology Change and Gender: Irrigated Agriculture and Peasant Women in Eastern Uganda

Zie Gariyo

Within the past two decades or so gender, and the women's question in particular, has taken on a new significance. In Uganda, and Africa in general, studies dealing with the women's question and the role played by women in agricultural production have been undertaken.[1] These studies have some common underlying assumptions. First, the advent of colonial rule and the penetration of capital enhanced the exploitation of women's labour. Second, women's role in capitalist agricultural production has not been sufficiently recognised. Third, capitalism consolidated the social and ideological structures that exacerbate the oppression of women. Sexual and social division of labour, control over productive resources, control over benefits from productive labour, and reconstruction of marriage and kinship institutions have enhanced the exploitation of women. Fourth, women's struggles against oppression and exploitation are primarily struggles against capitalist exploitation of women's labour and the resultant social structures, which have denied them a share of the productive resources.

On the one hand, the studies have recognised the extent to which women's productive and reproductive roles have been marginalized and on the other, the struggles by women to address the issues that enhance their exploitation and oppression. Thus scholarship has now been concerned with the dynamic relationships in society with women as actors and subjects of their own development as opposed to the earlier anthropological approaches that emphasised women's status and productivity in a passive perspective, i.e. their role in "traditional society". As Boserup remarks, "It is widely mistaken that such "traditional" systems are passed on from one generation to another without undergoing changes....".[2]

This chapter examines the triple relationship between women's labour, technology and irrigated agriculture with particular reference to peasant women's role in capitalist agricultural production. Of particular interest to this chapter is the extent of women's exploitation on the basis of the social and sexual division of labour as well as the environmental impact of irrigated agriculture on their living conditions. These relations are explain in relation to the social and ideological structures that enhance women's exploitation and the

M.A.M. Salih (ed.), Local Environmental Change and Society in Africa, 2nd Edition, 69–100.

varieties of economic, social and cultural struggles they wage to control their productive and labour resources.

The material presented here is based on the fieldwork I carried out among peasant women rice growers in Eastern Uganda, notably Iganga District where rice cultivation has been going on since the early 1920s albeit on a small scale. The collapse of the agricultural export cash crop sector in Uganda in the mid-1970s compelled the peasant communities in the region to look for alternative agricultural cash crops. This led to wetlands (swamps) reclamation in Eastern Uganda and the intensification of cultivation of flood rice, which had been hitherto carried out on a small scale. By the mid-1980s rice cultivation was the single most agricultural activity both for cash and consumption [Mwaka, 1989; Gariyo, 1991a]. The introduction of the Kibimba and Doho rice schemes merely enhanced the cultivation through the introduction of improved high yielding dwarf rice seed varieties (HYVS) and new agronomic practices like transplanting.

This research from which this chapter emanates was carried out among the peasant communities of Iganga District in Eastern Uganda. It is located in the wetland areas around Lake Kyoga, which covers 53,700 sq. km. which is a generally flat plateau, with a gentle dip towards the centre represented by the down warped area filled with lakes and swamps (Langlands, 1972 quoted in Mwaka, 1990). Iganga district, Bugweri and Bukooli have a large concentration of rice growers and were originally selected for our sample surveys. These however, were found to be inadequate. One village in Kigulu county, Nawankonge in Nawandala sub-county, and another one in Busiki county, Namutumba, were selected for household surveys. Farm surveys were carried out in Bukooli in the rice growing marshland areas of Idudi.

In Busiki, farm and household surveys were carried out around the areas of Namutumba and Busembatia, one of the largest concentrations of rice growing areas. 112 women and 30 men from 120 households were interviewed individually. The inclusion of 30 men in the sample was because of the complexities of the rural structure. In order to be allowed to talk to some of the women we had to talk to the men first. We followed up individual interviews of both the men and women with group interviews in order to ascertain general trends within peasant households on issues like gender relations, women's problems in agriculture, local politics, peasant taxation and peasant access to agricultural incomes.

Out of the 112 women interviewed, 85 (75.9%) were married women, 15 (13.4%) were widows, 6 (5.4%) were divorced and another 6 (5.4%) were single. Out of the 120 households selected in the sample, in 32 (26.6%) and 18 (15%) husbands had married second wives since they started cultivating rice. Out of the 120 households only 8 (6.6%) were relatively rich peasants employed as wage labourers during the rice cultivation season; 48 (40%) indicated that they employed casual (task rate) wage labourers for selected farm operations such as land clearance, weeding, transplanting etc. The bulk of the households rely on family labour for rice cultivation.

THE SETTING

The advent of colonialism and the penetration of commercial and industrial capitalism at the end of the 19th Century heralded in dynamic perspectives in agricultural production in Uganda which in turn affected the household social and labour requirements. The introduction of agricultural export cash crops greatly affected the existing social structures by introducing new forms of division of labour and led to intensification of women's labour. As Robertson and Berger observe:

Colonialism introduced new mechanisms of systematically extracting African wealth and new forms of patriarchal ideology and practice ... cash cropping was introduced mainly to men ... who also found more opportunities for wage labour. Pressured by the need to pay taxes, virtually everyone was brought into the nexus of the commercial economy while women retained the responsibility of feeding their families. The prevalence of male labour migration in many areas left them to do more agricultural work. Divisions were heightened which exacerbated inequality.[3]

Thus as men migrated to seek paid employment, as others were compelled to work on public projects viz. roads and the construction of public buildings, the bulk of the work of agricultural export crop production and the feeding of families fell on the women. Women's labour became a commodity to be exploited both by the men and the colonial state in pursuit of capital accumulation. The division of labour within the household, which had ensured that women and men performed different but complementary tasks in agriculture, was altered as women became responsible for all tasks [Tadria, 1985]. The changes in the division of labour where formerly men had performed the most physically demanding tasks so that women did not have to be involved, increased the intensity of women's labour as they took on both the most physically demanding tasks as well as tasks demanding longer periods of repetitive action (Gariyo, 1991a).

Carla Freeman, writing about a similar situation in Kenya, has argued that "the sexual division of labour created in very simple capitalist terms, a cheap labour force – because women sustained their families and reproduced them, they ensured the existence of a male labour pool, and enabled the industrialists to pay these workers wages below the subsistence level".[4] Although one might remark that Kenya was created as settler and industrial colonial state while Uganda remained a predominantly peasant economy, it suffices to note that the conditions of agricultural production had significant similarities. In both cases the colonial state mediated the exploitation of the workers and peasants.

The consolidation of the colonial state also consolidated the position of men by making educational opportunities and credit systems available to them, yet neglecting women who were primarily responsible for most of the agricultural production (M. Lovett, quoted in Parpart and Staudt, 1990). For instance in Buganda where 800 square miles of land was allocated as mailo (mile) land, i.e. on a freehold land tenure system, by the colonial regime to 1000 chiefs and other "notables", women were ignored. As Parpart and Staudt observe:

The state captured men far earlier than women but in exchange men acquired resources and used ascendance in state institutions to consolidate control over women. Nowhere is this clearer than in male control over land.[5]

The new production relations introduced by colonial rule enhanced the consolidation of men over the control of means of production. In other words although production was undertaken by the women, control over resources and incomes remained under the control of men. Bryceson and Mbilinyi for instance contend that:

State action enabled men to enter the money economy as ostensible producers of agricultural commodities, and ended up by denying women similar status despite the fact that their labour contributed heavily towards producing what was considered their husbands' crops.[6]

While women were responsible for most of the agricultural production, they did not control the income from the products of their labour. Cash income from coffee and cotton, the major export agricultural cash crops, were not under the control of the peasant women who produced these crops. Lack of control was, however, not limited to agricultural export cash crops only but also to other cash crops. Women were not allowed to sell crops like beans, peas, millet, maize, *et cetera* unless they obtained the permission of their "husband". If they sold them they were obliged to declare the proceeds to the "husbands" who would determine the amount to be used for household consumption goods like salt. Lastly, lack of control was even extended to domestic animals and their products, an important economic resource both in regard to their sale or the sale of their products.

In time lack of control was consolidated through a host of "traditions" and rules against women and for control over their labour. Women were not encouraged to take on paid labour but to stay at home which would keep them out of the cash sector of the economy (Tadria, 1985). Since independence, the conditions under which women live and work have not changed. What has changed in most cases is that the old cash economy based on cheap migrant labour, the export agricultural cash crops, has declined creating poorer living conditions for the rural communities in general and rural women in particular.

In Uganda the decline of the single season cotton sector during the mid-70s forced the peasants to look for alternative sources of cash incomes in agriculture. The intensification of a double season rice crop cultivation, which in the 1950s stood at a mere 2.5 tonnes, has therefore increased to the present 20 – 30 thousand tonnes annually. Thirdly, the collapse of the rural economy has brought about new changes in the rural economic and social structure both along gender and class lines. More women are now involved in agriculture, commerce, paid wage labour and so on, which means that they have taken on more and more responsibility and therefore exploitation and oppression.

THEORETICAL AND CONCEPTUAL ISSUES

Women's involvement in agricultural production is not a new phenomenon. Indeed a division of labour between the men and women existed before the advent of colonial rule. Tadria describes the division of labour based on age and gender as follows:

Men were responsible for providing shelter, clothing, meat and alcohol. Women produced most staple foods including potatoes, root crops, bananas, and various vegetables for sauce. They were responsible for food preparation and they manufactured mats, baskets, and other household goods (Tadria, 1985:36).

With the consolidation of the colonial state in directing and controlling agricultural production, new relations of production, i.e. control over the forces of production and over the means of production, changed tremendously. This is what M. Lovett says:

> The expansion of cash cropping and developing land shortages combined to bring about a rise in the value of land as property ... this translated into the increasing fragility of women's access to land as evidenced in part by emerging connects and gender struggles within the household over the allocation of land between men's and women's crops. While continuing to be responsible for food crop cultivation, women were expected to labour on their husbands' cash crop field as well. Women's labour burden accordingly was intensified and they progressively lost control over the products of their labour as well. ... women came to function in a role somewhat analogous to that of unpaid labourers on their husband's farms.[7]

Women's labour became a commodity to be bought and used or, as often happened, to be disposed off thereby breeding inequality among the sexes. With such inequality the regime of exploitation was expressed in the kind and the amount of work that women were expected to undertake as well as the length of time necessary to do such work both on and off the farm, i.e. the reproductive activity of women. These relationships were also expressed not only as the male monopoly over women's labour but also as a value free commodity (Barbara Brady, 1977).

Issues of women's consciousness, which form the character of women's organisations and struggles either to control their labour or transform the social and ideological structures will also be examined. This, however, will be done within the overall context of the rural economy, the dynamics of a capitalist economic structure and the means by which these have consolidated the poor conditions of the rural economy. The question that follows is whether changing the structures within which they operate can change women's working and living conditions. Our contention here is that the position of women in production cannot be understood unless we understand the dynamics of a capitalist production regime albeit a backward one.

GENDER AND CLASS: A CONTEXTUAL PERSPECTIVE

While predominantly dealing with gender, particularly the position of women in the production and reproduction continuum, this section of the chapter intends to examine the relationships that form gender and class. This is necessary because first, gender issues do not take place in a vacuum. Secondly, we recognise that the rural economic and social structure is not homogenous. Differential aspects and stratified categories among women do exist within which the social and economic relationships interact [Karen Sacks, 1982; Tadria, 1985].

It is therefore important to establish to what extent they have had **an** impact on women's consciousness and struggles. As Claire Robertson and Iris Berger contend "gender stratification is inextricably linked with class formation".[8] Women not only labour, consume and reproduce as part of their gender but also as part of the class to which they belong. And Mamdani has observed that "it is very important to recognise that production relations are gendered".[9] Therefore the organisation of labour, the consumption and the reproductive roles within the rural economic and social structure are different for poor and middle peasant households, rich peasant households or households which have access to property such as land and instruments of labour such as tractors, oxen and ploughs, hired labour and commercial capital (trade) which they use to acquire property.

Our contention is that in such a labouring regime, the organisation of labour seems to favour men. In poor and middle peasant households for instance women are directly engaged in productive labour both on and off the farm. In rich peasant households women undertake mainly supervisory roles especially of hired labour or engage in only those labouring activities such as preparing meals or washing clothes that are not physically demanding. This however, does not suggest that this type of labour liberates the women in this category. While it might be less oppressive and exploitative in terms of labour, it does not ensure automatic control over critical material and social resources such as land and cattle, or ensure the same leisure time as men.

TECHNOLOGY CHANGE AND PEASANT PRODUCTION

Swampland rice production was introduced to Ugandan agriculture by Indian traders in the 1920s mainly to sell to the expanding Asian and European communities. However, in the 1970s following the breakdown of transport, marketing, and facilities for processing cotton – formerly the most important source of income for peasant farmers in Iganga District – swampland rice growing in the basin was intensified [Nnabuguzi, 1993]. Consequently, rice production became the single most important cash crop for peasants in Eastern Uganda [Mwaka, 1990]. During the mid-1970s a rice experimentation and growing scheme covering 1387 acres and capable of producing about 5000 tonnes of paddy annyally, was constructed at Kibimba. In the mid-1980s another scheme was constructed as a rice popularisation and growing project at Doho in Tororo district capable of producing about 6000 tonnes of paddy annually.

The introduction of the two schemes introduced another aspect within rice and peasant production. Peasants were able to get access to high yielding variety seeds (HYVS) and to construct better water drainage channels. It also led to the adoption of new types of agronomic activities like transplanting where formerly sown broadcasting was used.

Although rice production like cotton production is also a single season crop, it introduced new agronomic requirements within the wet, muddy, clay swampland soils. It drew women firmly into agricultural cash cropping activities because, although both cotton and rice are labour intensive, women were only engaged in specific tasks such as weeding on the cotton farm, i.e. a clear division of labour in cotton cultivation. They now had to undertake all the tasks in rice cultivation.

Another important aspect is, that land distribution patterns within peasant communities changed. Swampland areas, which had formerly been reserved as dry season agricultural areas mainly for the cultivation of such food crops as potatoes, millet, beans, and grazing land for cattle and which were communally owned and utilised, took on a new significance. Peasants who owned land near swamp areas also claimed individual ownership of swampland. No longer were all peasants free to have access to the swampland areas as a communal right (Nnabugizi, 1993; Gariyo, 1992, Mwaka, 1990). A class of landlords had suddenly been created. This is what Nnabuguzi has to say;

In Busoga the complex system of succession that had emerged with the cash crop economy during the colonial period, underwent an important transformation. With the commercialisation of swampland and the accompanying increase in wage labour, land rights were being reshaped from below. The commercialisation of swampland brought forth new conflicts between systems of production, land use and land ownership...[10]

The creation of a minority class of landlords with control over swampland meant that land shortages and especially the most productive land became acute. No longer were women able to freely enter wetland areas during the dry season to cultivate food crops like millet, potatoes and beans, which are essential for household consumption. Some of these landlords do not cultivate rice themselves, but depend on the production of those who have leased pieces of land to pay them with a portion of the rice produced.

This could have serious implications for the rural and agrarian economy. Household food shortages and food availability become critical due to lack of cultivable land and due to lack of adequate incomes from rice production which would otherwise ensure that peasants are able to purchase food grain from other sources (Gariyo, 1991). Indeed, most peasants noted food shortages were becoming rampant as a result of rice cultivation and land shortages. Rice production, however, became important as a cash crop as a result of economic and social stagnation in Uganda during the mid-70s. With the failure of the state marketing boards to purchase the cotton crop from peasants, and the failure of the marketing boards to pay promptly for the cotton crop taken, compelled the peasants to look for alternative survival and cash income strategies (Mwaka, 1990; Gariyo, 1992; Nnabuguzi, 1993). The figures in table 4.1 below show the extent to which rice cultivation had by 1979 surpassed cotton production.

Production of rice intensified during the mid-1980s when peasants had completely abandoned cotton production in preference to rice cultivation. Rice production was also intensified because it enjoyed a steady price due to the ready internal markets from urban consumers as well as public institutions wary of heavy transport costs of other bulky foodstuffs like *matooke*. Presently rice production has not only overtaken production of cotton but also of many other agricultural cash crops such as maize, millet and cassava (Nnabuguzi, 1993)

GENDER, FORMS OF LABOUR AND THE LABOUR SITUATION IN RICE GROWING HOUSEHOLDS

The changes in labour and land relations have adversely affected the position of women. Rice production is first and foremost a labour intensive crop and women are engaged in all the rice cropping activities. The preparation of swampland involves the clearing of the papyrus forest by burning in the dry season and by slashing. The initial tillage operations are performed before flooding, and are followed by the boring of drainage system and transplanting.[11] Because there are no elaborate drainage systems such as those found at the Kibimba and Doho irrigation schemes, peasants enter the wet and muddy swamps to prepare the seed beds and till the swamp in preparation for planting the season's crop. Lack of water control and constant flooding occasionally washes away the new crop or the seedbed, forcing the peasants to sow more than once or missing the season's crop altogether.

Women, and especially women from poor and middle peasant households who cannot afford to hire expensive labour, are heavily engaged in almost all rice cropping activities. Both the men and the women wake up at around six o'clock in the morning during cropping time activities and work without a break before the hot sun becomes intense between 10.00 and 12.00 o'clock. Rice cropping activities include land tillage, seed-bed preparation, transplanting, weeding, bird scaring, harvesting, transporting from the field to the household compound, drying, winnowing, packing and transporting to market and selling.

Slashing of burned papyrus and land tillage using hoes take place during the dry season just before the rains fall to enable the peasants to turn the soil properly. Because of the hot conditions under which the peasants work, and the rudimentary implements used such as the hoe and the machete which require immense human power inputs, the amount of work undertaken remains limited. As one peasant reported:

I had planned to cultivate a piece of land of 20 *miigo* by 25 *miigo* (1 *muigo* = 3.7m) which would enable me to produce 60 bags for myself, but I could not manage. The cost of labour is so high that I could not afford to hire labour. I would have needed Shs. 38,000. I wanted to use my own labour… in the end I managed to cultivate a small plot, which would give me about 10 bags.[12]

Transplanting and weeding take place under wet and muddy conditions. Transplanting involves thrusting the young rice plants in bunches of four or five into the muddy soil using bare hands and which also involves upward and down movements of both back and hands. Weeding involves spending long hours of bending under the hot sun to pluck the weeds which could easily overwhelm the rice crop if not done properly. Because of the water, weeds grow very fast and weeding has to be done two to three times a season. This means that if it is not properly done the weeds could easily destroy the rice and makes the peasant loose the crop completely.

GENDER AND SEXUAL DIVISION OF LABOUR

The sexual division of labour is premised on the following assumptions. First, women perform different tasks from men. Secondly, women carry out less skilled and generally stereotyped tasks like transplanting, weeding, sowing inter-culture etc.[13] However, these two arguments cannot be generalised. For instance transplanting and weeding are not exclusively women's activities in rice production in Uganda. Both the men and women undertake them. In several circumstances you find men transplanting and weeding during the rice season. These activities are the most labour intensive ones. Transplanting takes a long time to do as it has to be done slowly and carefully so as not to destroy the young plants. And weeding has to be done more than once. Transplanting and weeding are the most crucial cropping activities and can be responsible for the amount of harvest one can expect. If they are not properly done, weeds can very easily destroy the rice crop. As one peasant put it: unless weeding is done effectively and in time it can very easily overwhelm you … this is what happened to me … because I had other things to do and could not keep up with cultivation schedules, weeds overwhelmed me. I had planned to do the job in the evening having failed to do it in the morning, and I did not. In the end weeds destroyed the rice crop I had cultivated.[14]

On the other hand, women are also engaged in physically demanding activities like slashing and tillage at the beginning of the planting season (table 4.1). Other activities in which women are heavily engaged during the rice cropping season include bird scaring, harvesting and transporting from the field to the household compound for drying. Bird scaring however, is mainly a child's activity and especially a task for young children who usually spend the whole day in the field. Bird scaring starts around 5.30 a.m. in the morning and ends at around 7.00 - 8.00 p.m. in the evening. It involves making a lot of noise and throwing mud balls to scare away the birds. Because some of the children go to school, women have to do it in the morning hours and the children in the afternoon hours. Because bird scaring is a full time activity it is common for children to stay away from school altogether during the period of bird scaring.[15]

Table 4.1: *Sexual division of labour within rice growing households*

Activity	Male	Female
1. Land preparation (burning, tillage)	M	F
2. Seed-bed preparation	M	F
3. Transplanting	M	F
4. Weeding	M	F
5. Birdscaring	M	F/child labour
6. Harvesting	M	F
7. Transporting	M	F
8. Threshing		F/child labour
9. Drying		F/child labour
10. Winnowing		F/child labour

Whereas peasant women are also involved in rice harvesting and transporting of paddy to the household compound, drying and winnowing of paddy are entirely women's activities. After harvesting women are responsible for threshing and drying the paddy until it is ready for sale. Before being sold however, rice has to be winnowed to remove the dust and waste matter left during the harvesting. Because winnowing is also entirely a woman's activity, if the woman cannot do it alone she will solicit the services of her "co-wives" and "sisters-in-law" on the understanding that she will return the favour next time she is asked.

Even the selling of rice is not entirely a man's job. There are two ways to sell rice. Because the amounts of rice are small, peasants do not find it attractive to take it to the milling machines, which are a distance away. Instead they sell the rice to the local traders who enter the villages at the beginning of the harvest season. The instrument of measurement used here is a tin. Most tins however, are not the same size and they are manipulated in such a way that the base is usually false. In this case therefore women are involved in the transaction. Because of the availability of numerous milling machines in the rice growing areas, some women also take their rice to the milling machines where the selling price is relatively better.

Women also do a very substantial amount of reproductive household chores. On top of being fully engaged in the cultivation of rice, women also have to produce other crops like maize, beans, millet, potatoes, and cassava. These are important for household food consumption and can be sold to obtain cash incomes. Peasant women also have to collect water, firewood, cook food, wash the children, wash their children's and their husband's clothes and generally clean around the household compound every day. This is on top of her other on-farm activities. Usually these activities are done single-handedly. Very few men are willing to take part. Out of 90 women respondents, only 14 (12.6%) indicated that the men would engage in one or more of the following activities such as collecting firewood and collecting water.

This implies that women's working days are longer than men's with fewer or no time for leisure. In the homes we visited during fieldwork for instance, we would talk to women while they are peeling matooke or potatoes, sorting the beans or doing any of the necessary household chores for the next meal. When we wanted to talk to the men, we would usually be offered a chair and have enough time to talk or we would find them at their local drinking places during the evening hours. Men find leisure time to mingle with their friends and have a drink, which women do not do.

In the cases where men were involved in performing some of the household chores, either the women were sickly or absent. Household chores include preparing the mid-day meal of cassava or potatoes with beans or green vegetables as the sauce, or merely boiling green whole maize to constitute a meal. Later during the day, preparations start to provide the evening meal of either millet, potatoes, maize meal and beans and occasionally meat. The most common mid-day meal is potatoes, cassava and beans or green vegetables. In some poor households a mid-day meal could comprise of only a piece of sugar cane or food left over from the evening meal. These households could only afford one meal a day, which is usually the evening meal.

Another aspect is related to hiring wage labour. During fieldwork for instance, we noted that hiring labour among poor and middle households was strictly confined to the cultivation of rice. But almost in all of them, it was a man's operations and the crop belonged to the man as the sole beneficiary of the income from the sale of rice. Men rent a small plot of swampland for rice cultivation from a landlord, hired labour is employed for tasks such as land preparation: transplanting; weeding; harvesting or threshing. The women take part in the harvesting, drying and winnowing of paddy. For men in this category, rice cultivation is the only profitable employment. When the harvest is good or is not destroyed by floods, rice production secures a reliable source of income from which they can pay taxes and meet other household expenses. Alternative forms of employment in the rural economy do not exist unless they also hire out their own labour as casual wage labour.

The use of child labour is prevalent in rice cultivation and is involved in both pre- and post-harvest activities. Child labour is mainly used for bird scaring during the rice season and for carrying loads of paddy to the drying area after harvest. Child labour is also involved in scooping paddy at the end of the day after the dry season. Some of these children who are still of a tender age are compelled to perform farm and household work whether they are going to school or not.

Child labour is not a new phenomenon in peasant agriculture. It is part of a peasant economy and indeed a backward capitalist economy and has been used in productive work for a long time. What is new, however, is that under other cash cropping regimes peasants would encourage their children to attend school since their participation in the cultivation of such crops as cotton or coffee was minimal. In fact the modest incomes from these crops were an incentive of peasants to invest in their children's education. With the present cropping regime children are instead withdrawn from school to undertake cropping activities such as bird scaring which last for more than a month. They are also involved during transplanting by uprooting the young rice plants because, according to the parents, "they do not use force which destroys the roots".

When children leave school, they begin to cultivate their own rice where they can be assured of modest incomes. After all, at school they were punished for poor performance resulting from missing out on classes for many days or months. Because the young people, and especially the boys, have access to incomes, they start families of their own by marrying other young people who are themselves victims of the same conditions. The implication here is that inadequate education poses a very serious problem for young women who often become pregnant when they are between the ages of 16 –20, undertake heavy cultivation of both agricultural and food crops, and look after the welfare of the household.

LABOUR INTENSITY IN RICE PRODUCTION

In this section we have attempted to determine the labour intensity for women in a rice production regime. The figures used are mere guesses but can generally reflect the actual situation in agriculture. In rice production, unlike other cropping activities, a working day starts at around 6.00 o'clock in the morning and lasts up to 10.00 – 12.00 o'clock. The reasons for starting very early are related to the wet and muddy conditions of the heavy clay swamp soils that become heavier as the cultivators tire out with the rising hot sun. Therefore the average working time in rice production in our case has been taken to be six hours a day. In other production regimes where the soils are soft and sandy loam it could go up to eight hours. Six hours in the swamp is a very long time and some cultivators end their working day by 9.00 o'clock in the morning.

The heavy workload of women makes their working day significantly different from the rest of the members of the family. Indeed we have already pointed out that most women are also involved in the same physically demanding cropping activities as men. They also do non-agricultural chores, and therefore work longer hours during the day. In table 4.2, we have attempted to calculate female labour intensity in rice production on the basis of an average of a 5 hours working day for the major cropping activities. The exceptions are activities such as seed-bed preparation (3 hours); harvesting (4 hours); threshing (2 hours); drying (2 hours) and winnowing (3 hours). These are activities, which do not require a full working day because they are either done after other agricultural work in the evenings or by setting aside one single day for a particular task such as winnowing which must be done before the rice is sold.

Table 4.2: *Labour intensity for women in rice farming activities*

Operation	Labour Time in Hours/acre/household		
	Total Labour	Female Labour	% of total time used for all operations
Land Preparation (burning, tillage)	420	370	12.7
Seed-bed preparation	30	30	1.0
Transplanting	300	250	8.6
Weeding	630	525	18.0
Bird scaring	1260	720	24.7
Harvesting	160	160	5.5
Transporting	-	-	-
Threshing	60	60	2.1
Drying	30	30	1.0
Winnowing	27	27	0.9
Total	2917	2172	74.5

Source: Field work data.

The calculations in table 4.2 are premised on the assumption that each household has five working people, two adults and three grown up children and that each individual does 6 hours of work per day. Women however, do not spend the full working time in the field since they have to take time off to prepare a meal for the household or breast-feed the baby. The figures in the table indicate that total amount of time that women spend on agricultural tasks is 74.5% depicting their heavy involvement in agricultural work.

Table 4.2 also indicates that women's work includes heavier activities such as land preparation (12.7%); transplanting (8.6); weeding (18.0); and bird scaring (24.7%). Weeding is sometimes done two to three times a season. If we take each weeding to take a week per acre, then a household will require three to four weeks to do all the season's weeding. Birds caring has a higher percentage at 24.7 because it takes almost 30 – 40 days before harvesting so it sometimes starts much earlier to prevent the birds sucking the milky seedlings of the young plants as they are flowering. Secondly, birds caring needs an average of 14 hours a day because it starts around 5.30 a.m. in the morning and lasts almost up to 7.00 – 8.00 o'clock in the evening.

Bird scaring is primarily children's activity and indeed during this time most children do not attend schools but have to stay behind to tend to the rice. Many women however, are also involved in bird scaring activities mostly between 5.30 a.m. and 2.00 p.m. when the children return from school and take over. Although the percentage of labour input in activities such as threshing, drying, and winnowing is low and looks rather negligible, they do not reflect the fact that they are exclusively female activities. Women spread the rice to dry in the sun before going to the fields and scoop up the partly dried rice for storage at the end of the day. Male labour time in rice production however, is lower than female labour time. This is because we know that male labour is only available for activities such as land preparation, seedbed preparation, transplanting, bird scaring, harvesting, and transplanting. Male labour involvement in activities such as threshing, drying and winnowing remains limited.

GENDER, ACCESS TO IMPLEMENTS OF LABOUR AND AGRICULTURAL PRODUCTION

We have already noted that the most immediate implements in peasant agriculture are the hoe and the machete. Peasant agriculture however, also has access to other instruments of labour like the oxen and ploughs. Occasionally these are hired for performing agronomic activities like ploughing which is one of the most physically demanding activities in agriculture. Because of the high cost of hiring ox-ploughs, they are not commonly used in swamp rice cultivation as most peasant households cannot afford the cost of hiring an ox-plough for cultivation. Besides, the soft muddy swamp soils are not convenient for ox-plough cultivation.

The use of ox-ploughs and even tractors was common in cotton cultivation until the mid-1970s, because until then the implements of labour were partly subsidised with owners of oxen buying ploughs at reduced prices and hiring in tractors from the Department of Agricultural Mechanisation of the Ministry of Agriculture. Besides, the costs of hiring an ox-plough was fairly low compared to incomes from cotton. This enabled peasants to use either a tractor or a plough.

Some peasants reduced the costs of hiring tractors and ploughs by either buying their own ploughs and paying for the costs of oxen or keeping the oxen and hiring a plough. Both the plough and oxen were hired out for extra income for the owners after performing cultivation tasks at the household farm.

The division of labour during this labouring regime was such that the men cleared the land using slashers, machetes, and axes to remove large wood stumps and the overgrown bush around the planting area. The ox-plough and the tractors were used to turn the soil once or twice before planting. Women were involved in cotton cultivation when it was time for weeding and picking. The importance of such cropping programmes was that as the men were busy preparing the cotton fields, the women would be busy preparing fields or planting other crops that were used both for cash and household consumption.

Mechanical cultivation however, was only favourable for men to lessen their work during the cultivation processes. Either the tractors or the ox-ploughs performed the most physically demanding activities. The collapse of the system and therefore the use of mechanical implements together with the intensification of paddy cultivation in Busoga eliminated the advantages for men and drew women more firmly into agricultural processes. Women now entered the cultivation process using just a hoe and a machete at the very beginning and not just for two or three cropping activities as used to be the case during the cultivation of cotton. Moreover even hoes and machetes are now too expensive to purchases after the withdrawal of subsidies for agricultural implements from the state.

Yet women still remained largely responsible for providing for the welfare of the household and were therefore compelled to cultivate other crops for household consumption thereby increasing the amount of agricultural work they had to perform. This meant that the introduction of labour saving techniques in agriculture would not necessarily favour women since these techniques tend to be monopolised by men. As Prem Choudhry has noted in the case of the green revolution in India "...mechanisation ...lightened the workload of men...the workload of women has only been recast, not reduced." [16] This is true even in a situation like ours where women's workloads have increased with the introduction of crops such as rice which purportedly increase the incomes of peasants.

GENDER, COOPERATIVE LABOUR AND AGRICULTURAL PRODUCTION

However, the exploitative labour regime described above has made women aware of the need to divide their labour between labour intensive activities (rice) and activities necessary for household welfare (food). In order to cope with the non-agricultural household tasks and the labour requirements on swampland rice cultivation, sections of peasant women have now organised themselves into labouring groups of two or more women. Women from different households come together on a number of days and work on a member's household plot. Here women grow crops like cassava, beans, millet, maize, et cetera. Cooperative labour, which would shorten the number of days worked on each task if it were used, is, however, limited to the cultivation of crops other than rice.

A group of four women organised in a cooperative labour group informed us of the reasons for this trend: We women have no rice gardens of our own. We work in our husbands' gardens and during our free time we are usually involved in mat making and making table clothes. We have no commercial (*sic*) crops …the crops we grow like potatoes and cassava are mainly for food. We can only sell if there is a surplus above the family demand.[17]

Although these women have indicated that the food crops produced are not for commercial purposes but for household consumption, most households admitted that a portion of the crops they cultivate is sold for cash. Other crops such as maize and cassava are the major cash generating crops in the region. Most of the households in our survey indicated that at least a portion of each of the food crops grown such as cassava, beans, maize, ground-nuts, soya, and millet were also sold off to obtain cash incomes. In fact in some cases crops such as soya, which is not a popular food crop is cultivated to obtain cash income. Besides, because of storage and processing problems such crops like cassava and maize, which are widely grown in the region, are actually sold off to avoid wastage and rotting due to weevils, high moisture content, rodents, and occasionally petty theft in the villages.

Despite the intensity and long hours spent working in agriculture and the negative implications it has for women's health (see Mwangi in this volume), they have not managed to institute a cooperative spirit to solve labour bottlenecks. Cooperative labour has mainly been used as a survival strategy between a labour intensive regime like rice with its high labour requirements and the needs of households in which a cash economy has to compete with subsistence needs. The labour intensive demands of rice cultivation undermine the households' abilities to provide their own food requirements however, inadequate the quantities may be.

The use of cooperative labour is so popular that even men have their own co-operative labour groups where they concentrate on the production of commercial agricultural crops. Unlike the women's cooperative, where cooperative labour is mainly for the cultivation of crops for household consumption, men's cooperative labour is largely for the cultivation of crops for cash incomes. Most of the time, cooperative labour is for the cultivation of crops like rice, maize and cassava, which are largely cash crops. A member is only entitled to an amount, for instance one tin, which he can take home for consumption. Men's cooperative labour is also used for other income generating activities like lumbering and brick making which women are not allowed to be engage in for fear of competition.

TECHNOLOGICAL CHANGE AND PRODUCTION RELATIONS: WOMEN AND ACCESS TO LAND RESOURCES

The last decade has witnessed a proliferation of land tenure studies in Uganda.[18] However, these studies are in most cases too general and do not deal specifically with women and the land question. However, a few references have been made to women and land relations within a peasant and rural economy. These studies note that both the fragmentation of land due to inheritance practices within poor and middle class peasant households further marginalizes women, since women: 1) are not entitled to inherit land not even their parents' or husbands' homes. Even with the death of their husbands, women remain occupants of the land in trust of the children and male children especially. It is not uncommon to find widows without sons being disinherited after the death of their husbands. 2) Women do not own the land they till at their husbands' place but can use it as long as they stay at the matrimonial home. 3) Women can only own land if they purchase their own plots.

Women in Busoga like in any other male dominated society in Uganda do not own land. Married women are only allowed to use land as long as they remain married to their husbands. Land rights for women remain at their parents' homes in case they do not get married or in case their marriages break down. Land rights however, even here entitle the women to use the land to produce for her subsistence and for earning incomes to sustain her non-food needs. The land the woman cultivates at her parents home cannot be bequeathed to her children after her death.

In a few cases however, non-married women with children could gain the sympathy of her parents who would allocate a piece of land to her for the construction of a house, for the cultivation of crops for her household consumption and the cultivation of crops for sale to obtain cash incomes for sustaining her own household. For instance out of 22 women (table 4.3) who claimed to own land, only one married woman claimed to have been given a piece of land at her parents place, 3 single women had bought pieces of land. Three elderly women who had separated from their husbands after they had married other younger women, had received land from their brothers whom they were now staying with.

Table 4.3: *Women's land ownership*

Land category	Number of Households		
Status of Women	Inherited or Donated	Bought	Total
Married	1	-	1
Single/Separated	3	-	1
Widowed	3	3	6
Total	16	6	22

Source: Fieldwork Data.

Out of the 15 widows in the sample, 12 were still occupying their husbands land and 3 had managed to acquire land after their older working sons had bought them each a piece of land on which to settle with the younger children. In the event of death, however, the rights of ownership revert to the son who bought the land. Women's land ownership rights do not exist in reality except as destitute members of the community.

GENDER AND IDEOLOGY IN AGRARIAN ECONOMIES

The denial of women's access to the most important productive resource in an agrarian setting is both related to the conditions of the rural economy, i.e. capitalist working conditions, and to the ideological expressions under which such conditions are reproduced. During the course of the field study for instance, we asked women why they did not own land either at their husbands' places or at the parents' places. The answers were varied. Some replied, "Because we are women". Others indicated that it was because of the "mobile character of their existence" in that they had to leave their parents to go and live with their husbands. The parents used the excuse that because the women's life is a "nomadic type" of existence, they cannot own land.

Most men interviewed also answered, "Traditionally women are not supposed to own land". Thus "tradition" is invoked to deny women access to vital economic resources such as land. This however, does not explain why most men were not willing to allow their "wives" to rent plots of land in swamp land to grow rice of their own. Indeed landlords informed us that they couldn't allow married women to rent land from them without the permission of their husbands. Only young or unmarried women did not have such restrictions.

The most plausible reason would therefore seem to be economical. Men fear women having their own resource base because this would increase the competition between two equal partners in terms of decision making, both at the level of the household and at the level of the farm. Indeed, studies that have attempted an explanation for these relations have indicated that there is a degree of independence in decision making by women who are economically well-off than those who are not (Tadria, 1985; Akinyi Nzioki, 1991). Because in an agrarian economy women are primarily regarded as labour power and not as social partners, this would complicate issues of labour supervision and control.

Few men take part in household work. With a high degree of independence in decision-making, some of the traditional practices would be questioned. Indeed in another study that we undertook earlier, it was noted that women's decision-making and consciousness increased with economic power. Women were beginning to question some of the traditional practices and especially those that attempt to keep their labour under the control of men. After women had accumulated their own money they separated from their husbands to go and set up their own small businesses. Others protested against their husbands taking on another wife.[19]

Some men also conceded that the introduction of rice schemes had altered some of the traditional practices like polygamy and that while the rates of youth marriages had increased, the rates of separations were also on the increase. The reason given was that women were beginning to obtain economic power from the incomes from the cultivation of rice, which enabled them to set up their own households. More women were being drawn into small trade in rice leaving their husbands behind to look after the households, which most of them resented.

The oppressive tendencies in a peasant economy are largely manifested and compounded through the institution of marriage and bride price. Women's marriage is both encouraged and celebrated on two grounds: First, as a social occasion and second, as a source of wealth for some households. We would like to discuss this aspect from an economic perspective. Recent analyses of social trends in Busoga indicate that early marriages, i.e. marriages between young people, are becoming rampant due to accessibility to incomes from rice. Parents encourage these early marriages because they are also ensured some form of payment either in cash or cattle (Mwaka, 1990; Gariyo, 1991) (a) (b); Nnabuguzi, 1993). This trend has given rise to a new phenomenon. Young people who were heavily engaged in rice cultivation on the basis of cooperative labour with other youths withdrew their labour from the cooperative and ceased to grow rice directly. The young women now undertook much of the work. The young men turned to enjoy their leisure instead[20] Young men therefore look at marriage as a source of agricultural labour on the one hand. On the other, they look at it as their own liberation from the strenuous manual labour that they leave to women to perform.

WOMEN'S ACCESS TO CREDIT

It has been observed elsewhere that women do not have access to formal credit (Akinyi Nzioki, 1991). This is largely because most banks need security which women lack since they do not control land which banks need as collateral. Even where credit facilities for peasant communities are provided, very few women are selected to benefit. Lack of access to credit for women ignores the fact that women are the people directly responsible for agricultural production. In Uganda for instance, there is a Rural Farmers Credit Scheme (RFCS) administered through the local Uganda Commercial Bank. With its network of branches it was expected to bring credit facilities closer to the peasants.

During the course of the fieldwork, only two peasant groups from the sample had ever benefited from the credit facility made available through RFCS.[21] Peasants in general are at a disadvantage in securing such facilities. First, formal bank loans are available only to rich peasants and capitalist agricultural producers. They are the ones who have access to land titles. Middle and poor peasant cultivators do not have access to land titles because they are expensive to obtain and the procedures are rigorous. Secondly, large portions of the credit from the rural farmers' scheme are not provided in cash but in materials. These compulsory materials, which include items such as wheelbarrows and agro-chemicals are not very attractive to peasants. They need the cash to buy hoes, *pangas*, harvesting knives *et cetera* or for hiring wage labour. Thirdly, these items are highly priced and yet repayment is expected in cash after harvest. The vagaries of nature including floods, birds, rodents and storms occasionally destroy the crop rendering the peasant unable to pay and heavily indebted.[22]

Lack of formal access to credit however, does not mean that women have no access to credit at all. Indeed, women have either formed their own informal credit networks or belong to those which already exist. There were several types of women credit networks identified during the course of the fieldwork. First, there are women's clubs. These operate like savings and cooperative societies, which require members to deposit a specific amount of money with the club and borrow against aggregate deposits for a specific period. Repayment includes a nominal interest. These clubs are, however, usually dominated by the women of local bureaucracies like teachers, agricultural extension officers, local government (chiefs) officers, traders *et cetera*. Very few attract the poor peasant women.[23] Secondly, there are other informal village structures like *Munno Mukabi* (distress assistance), which were established by the communities to assist people in distress such as meeting burial costs. Some of these have now been institutionalised to provide short-term credit to those in need to solve their own immediate problems. Because men dominate these structures however, very few women are able to obtain credit facilities from them.

WORKING AND LIVING CONDITIONS OF WOMEN

Rural women do much more work both at the level of the farm and at the level of the household. With improved incomes from agriculture due to rice production, one would expect that an improvement in living and working conditions of rural women would be noticeable. Except for peasant households which rent out land, are involved in small trade, or are large rice producers, the majority of the peasants live a distressing existence. In other words, different peasant classes benefit differently from technological change. We shall however, have the occasion to expound on the class related benefits in a larger study of peasant agrarian and production systems at a later stage.[24]

In most cases technological change in agriculture, rather than liberating women's labour, has instead intensified it. We have also already noted that women's labour has become commercialised so that it can be bought and sold. The conditions of work of women have a bearing on their conditions of living. Because much of the day is spent in agricultural and household work, they have few hours of rest. Because their labour time in agricultural work is not paid for they have no access to such necessary consumption goods like sugar, soap, salt, milk, cooking oil, meat, *et cetera*.

During our fieldwork we carried out a survey to determine the rate of purchase and consumption of non-agricultural commodities. From the 90 households who responded to the question our results were as follows (table 4.4).

From the figures given above some interesting observations can be obtained. Although the amounts purchased are not given in the table (we know that amounts purchased per commodity do not exceed one kilogram, litre or piece), the field results show that most peasant households do not frequently have access to adequate amounts of non-agricultural commodities. Indeed, some families never drink milk or buy meat. Most households drink tea and eat porridge without sugar. Men usually cope by buying sugared tea and a few pieces of roasted meat at the nearest evening market during their drinking hours. This situation is not surprising given the costs of meat, milk and sugar. At the cost of Shs. 1200 (equivalent of 1 US dollar) per kg of meat, most peasant households which are also large families, find it prohibitive to buy meat as regularly as possible. According to male responses at an evening drinking place, "the costs of meat cannot allow us to purchase meat for home consumption given our big households". This does not explain, however, why men prefer buying bicycles and radios after rice harvest or pay bride price for a second or third wife.[25] This raises serious questions of priority purchases within peasant households. As some scholars have commented, "the income of the cash crops is controlled by men, who do not spend the money on the necessities of life".[26]

Table 4.4: *Peasant's consumption levels*

Times per month	0 – 2	3 – 5	over 5
Commodity	Number of households		
1. Salt	56	34	-
2. Soap	48	26	16
3. Paraffin	56	28	6
4. Sugar	66	16	8
5. Meat	51	29	9
6. Milk	64	14	12

Source: Fieldwork Data

Other factors however, affect women's conditions of living and of working. Apart from hard farm and household work, women have to contend with issues of pregnancy and poor medical facilities in the rural areas. During fieldwork for instance, most women contended that pregnancy is one of the most burdensome issues, which impinge on their health and increases their workload. Pregnant women continue working in the fields until a few days or even hours before delivery. One or two weeks after delivery and before they have been able to regain enough strength they resume agricultural work. Household work however, is a continuous process, which they resume two or three days after delivery.

As most women pointed out, "we peasant women never rest or retire from agricultural and household work until death". With such trends almost all of them complained of being constantly sick with malaria, abdominal pains, backaches headaches and nervous diseases. Poor housing and sanitation conditions compound such diseases. Most peasant homes are mud, wattle and grass thatched houses. They lack adequate room and ventilation, as most of the houses are merely one or two tiny bedrooms. The beds are just simple and often with grass stuffed clothes used in place of mattresses. Very few households can afford a sponge mattress. The floors are merely covered with cow dung that becomes dusty after a few days necessitating replacement.

Under such circumstances peasant women are expected to perform their duties and live a normal life. Irrigated rice production, which should have ensured reasonably modest incomes that could have changed some of the living conditions, have merely compounded them- first, by increasing their labour and secondly, by denying them basic necessities.

TECHNOLOGICAL CHANGES, CHANGES IN THE RURAL AND AGRARIAN ECONOMY AND GENDER

From the foregoing discussion and from studies carried out elsewhere the structure of the rural and agrarian economy is premised on some assumptions.[27] First, it is a "primitive" capitalist economy with extremely exploitative production relations. Second, any changes in production and agricultural regimes can only build on the existing backward social structures of relying on cheap female and child labour. Third, the rural and agrarian economy is a contradictory setting both as centre of production and reproduction. The agrarian structure in Uganda falls under such a category. The reliance on cheap family labour means that production of a marketable surplus and production in general cannot expand. This limits the levels of income, consumption and the changes in household reproductive roles, which are associated with different levels of socio-economic development. Economic resources accumulating at the level of the household remain limited thereby also limiting the ability to satisfy social and consumption needs.

In a situation like this, and with the introduction of new productive regimes in agriculture, it is expected that there would be increases in wage labour employment and in particular farm labour. This is important because on the one hand, it would provide an income base for rural people, which would improve their living conditions. On the other, women's productive labour would now be available to provide adequate household consumption goods thereby reducing their workload, which should improve their own welfare. A few of the women would themselves constitute part of wage labour so that they can earn their own incomes and have an independent income base rather than relying on the men most of whom are themselves poor.

From our field work findings however, very few households are able to hire agricultural wage labour and where this does takes place, it remains casual labour limited to a few agronomic practices like land preparation and harvesting. The most labour intensive practices like transplanting, weeding and bird scaring are done using family labour.[28] Only a few households are able to hire agricultural wage labour on a full-time basis during the planting season.[29] Our findings also revealed that such casual labour is confined almost exclusively to the cultivation of rice and not any other agricultural crop. As such agricultural wage labour is not available for the cultivation of primary food crops, which are the main occupation for women in the rural economy.

Limiting women's labour input in agriculture or providing them with a source of income through wage labour is not sufficient to improve their lot. If the amount of off-farm work remains high or levels of income through rural wage labour remain inadequate, this might itself compound their situation. We are conscious of the fact that issues of liberating women's labour and improving their welfare are more complex than having to do less work in agriculture and having access to wage incomes. Our contention however, is that the struggle against backward production relations in agriculture is part of the wider struggle to liberate the labour of women and rural women in particular.

With the intensification of rice production, new changes are beginning to become noticeable. Rural wage labour is beginning to manifest itself as a source of wage income for the rural people although only men are free to participate so far. Sections of the middle and rich peasants have been able to construct modest houses made of bricks and with iron roofs and some of the women have been able to earn modest incomes from their own from rice cultivation.

In spite of these few trends, rice production has not noticeably altered the agrarian and social structures. The degree of backwardness of the rural economy remains high, dietary conditions and conditions of living of most peasant household remain poor. Access to health facilities, clean drinking water, good roads and proper schools remains non-existent. During fieldwork peasants kept asking us, "We are forced to pay high taxes and yet we do not have drugs in the dispensary. Our schools are rotting, and we do not have clean drinking water. Where does the money that we pay go to?"

Indeed, because of rice production, peasants reported that the rate of taxation had increased many times. The chief and the local resistance committees were the most unpopular people because of their high handedness in levying taxes on poor peasants. The impact of taxation is also great because, while formerly the guarantees of prices for cash crops like coffee and cotton slightly cushioned the peasant, this is presently no longer the case. The price structure of rice is different from that of coffee or cotton. While in most cases the market price is higher per acre cultivated compared to the other crops like millet and cassava and should therefore guarantee a modest income to the peasants, the operations of the market have a different impact.

Let **me** explain. During fieldwork we found out that most rice cultivators do not wait until after harvesting to sell their rice crops. Because of their immediate demands for cash, they are compelled to sell their crop at pre-harvest prices (*Kutunda Makoola*), which are almost five to ten times lower than the post-harvest prices.

How does this happen? A peasant rice cultivator who has problems to solve and needs immediate cash approaches a rice trader to borrow some cash. They agree that the peasant rice cultivator undertakes to repay the trader in the form of rice (and not the amount of cash borrowed) after the harvest. The price agreed however, is the price per tin of paddy at the time of taking the loan and not the price at the time of harvest.

We found out that the peasants usually sell their paddy rice at between Shs. 500 – 1000/= per tin at the pre-harvest price. After the harvest the price is usually between Shs. 3500 and Shs. 7000 per tin depending on the time of harvest, which is higher at the beginning and at end of harvest. The losses of the peasant are so immense that most of them have become destitute as a result of rice cultivation instead of alleviating their social and economic problems. This is what one peasant told us:

I had a problem. My children were sick and I needed money urgently. I contacted one of the buyers. My paddy was still young; I had not even started weeding. I was given Shs. 500 per tin. After harvest the price was Shs. 4500 – 6000/= per tin. I harvested 50 tins but I had sold 40 tins at Shs. 500 per tin and was only left with 10 tins.[30]

A situation like the one above is not helped by the fact that the local political and administrative machinery sides with traders. In fact, the traders have become too clever because they make the peasants sign written agreements after a few of them had reneged on unwritten agreements. Their relationships with the landlords, the owners of the swamps, also compound peasant exploitation. Most landlords do not take cash for the land they rent out to peasant cultivators. Others take cash plus paddy. For instance we found that some landlords demand between Shs. 2000 – 5000/= per plot (a plot is slightly smaller that an acre) plus one or two bags of rice. The 2000/= is charged for all new plots and is like an entrance fee. Most of the land cleared is forest or very thick bush but some peasants complained that a landlord "...can chase you away after only one season".

Landlords however, benefit immensely because they collect paddy from many peasants who rent land from them. Indeed, one landlord with about 8 acres of swampland who started cultivating rice in 1935 has stopped since the early 1980s. Instead he lends out most of the land to 7 – 10 tenants and collects 2 bags per acre each or an average of 10 bags per season. Such a landlord has an average income of Shs. 300,000 – 400,000 (about US$ 400.-) per season.

What does such a situation imply for women? Women are a specific part of the rural social structure, which has overtime institutionalised their exploitation and oppression. Because the new changes, i.e. the introduction of new forms of cropping patterns and new forms of accumulation, have marginalized whole communities, this compounds the marginalization of women. The introduction of new labour demands in agriculture for instance, does not change the character of women's labour but instead intensifies women's farm labour input. It does not radically reduce the demands for female labour in agriculture or the absorption of wage labour in. Instead, institutions like marriage are given completely new meanings. Men look at women's labour as a commodity and because of the relatively modest incomes from rice cultivation they take on more wives to increase production.

The most dominant feature of rural peasant women is that they provide farm and household labour. In fact they are looked upon as such. This is in contrast to women in the other sectors of the economy, the urban or semi-urban sector where women also take part in commercial activities as well as taking time off to relax or paying for some aspects of farm and household labour. This is what Tadria's study in the two villages of Segukku and Nddejje (just a few miles away from Kampala, the main commercial capital of Uganda) implies. While the women in the two villages near the city were able to take part in other non-agricultural economic activities and interacted in the daily food markets either as sellers or buyers, many women in the villages of Iganga District do not have such an opportunity.

The food and grain markets in Iganga do not have similar advantages as those near Kampala. The small village grain and food traders do not pay market prices for the crops they purchase from peasant women; the prices are usually determined by the same traders. There are only a few traders but there are numerous peasants selling small amounts of grain which renders their bargaining power very limited. The real competition is not between the traders but between the peasant women themselves who are willing to take any amount of money as long as it is paid in cash.

WOMEN'S CONSCIOUSNESS AND STRUGGLES TO CHANGE THEIR LIVING AND WORKING CONDITIONS

The impression created by the preceding discussion would tend to indicate that women are docile, apathetic and unable to change their conditions of working and living. This is partly true as by now we are aware that the social, economic and ideological structures in the rural and agrarian economy have for a long time remained intact in spite of the introduction of advanced technological and agronomic practices. Women remain confined to the home as the arena of production, reproduction and consumption. Their labour largely remains the source of agricultural production and their access to a share of agricultural income remains limited. Women remain unequally subjected to the structures of the society, which regard their labour as a free good to be exploited and disposed of at any time. Although a measure of protest by some women themselves has occurred, this remains spontaneous and limited. Women have now internalised the ideologies that have compound their inequality to the extent that many of them are unwilling to join those who are agitating for change.

Our own findings tend to suggest that women and in particular rural peasant women in Uganda have not responded collectively to address the issues of inequality between sexes and unequal relations of production. The actions undertaken are largely individual responses to specific situations. For instance, although we have already indicated that a few women have been able to acquire land, the majority of them have received land in the form of donations from their parents, after their husbands' deaths or because of their unmarried status. The response of the majority who are married indicates that women's consciousness towards property or capital accumulation is still low. The right to capital accumulation is one important aspect of women's emancipation but as long the social structures in society still emphasise inequality between men and women, this emancipation can only remain limited.

Within scholarship, it is suggested that women's empowerment and the transformation of inequalities denote control over their own labour both as paid labour and for household reproduction (Tadria, 1985; Elson, 1991; Gariyo, 1992). These views which are generally economistic have been used to "naturalise inequality and to obviate dissent within as well as beyond the household".[31] Most analyses which emphasise the aspects of women's control over their labour, control over their income, and access to markets as a liberating experience are removed from the rural and peasant setting to an agricultural plantation or semi-plantation and urban setting.

An earlier study on the Kibimba and Doho rice schemes in Eastern Uganda revealed that, whereas the working conditions of the majority of the women were poor, some women now had access to modest incomes through being allocated or leasing rice plots at the schemes. Because of these incomes it was also revealed that a new consciousness was beginning to set in as some women were able to leave their polygamous husbands and set themselves up independently either in business or through the purchase of land as well as constructing their own houses (Gariyo, 1991). However, these are individual achievements. Most rural peasant women do not have opportunities which guarantee them an income or property accumulation. The attitude of men towards women who set up their own households is rather derogatory. Such single woman headed households are known as Nakyeyombekaire in Lusoga or to put it more appropriately "the women who have built their own houses".[32]

This characterisation of women is a result of a society that has always taken women as incapable of undertaking activities independent of men. These attitudes are reinforced by the socialisation (ideological) processes within households, which prepare women not as economic actors but as material for marriage. These attitudes and identities deprive women of the necessary dynamic to carry on the struggle to liberate their labour, accumulate property, and participate in the commodity markets thereby depriving them of some economic power. They remain politically apathetic to the extent that they do not relate their participation in civic and social organisations as part of their larger struggle to understand the forces in society that compound their oppression.

During fieldwork we were interested in finding out the frequency of women's participation in Resistance Council (RCS) meetings and related activities.[33] Our findings revealed that most women had not attended any meetings of the council for many months after the elections. Others did not know where such meeting took place and who their women leader was. Yet most of the women indicated that they frequently attended meetings called by the local church pastor or belonged to the church sponsored associations such as the "Lay Apostles" which provide free labour to church related projects. Others confirmed that their husbands do not encourage them to attend the meetings because according to the men "there are no women issues to discuss" thereby confirming the view that men are not willing to let the women think for themselves.

As one woman put it, "We do not attend RC meetings because our husbands are not in favour of our attending. I used to attend RC meetings before I got married but since then I have not attended any meeting. I was not even told when there were elections, although I could even manage the post of chairperson if I were elected, but I am afraid it might conflict with my position as a wife because it might give me less time to attend to my husband".[34]

The secretary for women's affairs in the Resistance Committee II (the parish level) complained that men refuse to allow their wives to attend the seminars because according to some "...they are sometimes taught things such as family planning which are detrimental to the running of the home". Most of these men were not willing to discuss the fact that seminars for women are multidimensional in that they discuss family planning as well as nutrition, health, income generating projects, and agriculture which are at the core of women's welfare. Some few women however, were active in the RCS and had been elected to some of the high posts such as vice-chairperson on the village committee.

The views cited above are characteristic of rural, social, economic conditions that do not tap the existing initiatives of members of a community. Indeed, during fieldwork, we noted that many women were coming together to form groups and organisations to undertake some socio-economic activities whether the men approved or not. But as we have already indicated these groups are confined to the more educated rural women. Some women groups use women to solicit financial support from the numerous foreign funding agencies in the country, or from the politicians looking for a social base in the countryside. In fact many of the groups are formed at the instigation of the politicians who promise to offer them assistance if they form such a group. Almost all the groups we encountered fell under this category.

CONCLUSION AND POLICY IMPLICATIONS

The analysis of the working conditions of peasant women is not exclusively the analysis of the agrarian and rural social and economic structure. Hence the solutions to the problems of peasant women are not isolated from the solutions to the existing problems of other categories of peasants. Woman empowerment must take into account the social, economic, ideological and cultural structures in society.

Although woman empowerment requires that women should assume control over land and labour and income, they should equally enjoy the freedom of access to participate in commodity markets. Even here women are disadvantaged. In a sense, women's daily struggles to overcome these constraints are immersed in social, ideological and cultural structures that have historically formed the core values of women's exploitation and oppression. We would like to note that ideological and cultural structures are largely responsible for denying the recognition of women's role and contribution in agricultural production. This implies that the efforts by women themselves, which seek to address the issues of accumulation through acquisition of property, cooperative and hired wage labour, and participation in agricultural trade, should be complimented and concretised with understanding of the forces in society that have an interest in the subjection of women.

This chapter indicates that powerful forces in society strive to maintain the status quo confronting women. The policy implications of this are manifold: Firstly, more analytical research with the objective of understanding the structures and forces in the rural and agrarian economy that constrain the desired improvements in the poor living and working conditions of women. The formation of rural peasant women's and rural women's organisations for collective group actions, their participation in trade unions and local political structures, are deliberate political struggles necessitated by the grip on reality of the rural poor.

This chapter also shows that the distribution of the productive resources such as land, and facilities such as credit and technology are not gender neutral in any society. Therefore the restructuring of the productive endowments and facilities is a prerequisite to women's empowerment in order to reap the fruits of their labour and the full utilisation of the productive resources from which they have been excluded for so long. In essence, women's reclamation of the local environment, which provides these productive resources on which they survive, is a prerequisite to a truly empowering agrarian as well as technological progress.

ENDNOTES

1. See for instance studies by H. M. K Tadria (1985) *Changing Economic and Gender Patterns among the Peasants of Ndejje and Seguku in Uganda* PHD Thesis, University of Minnesota; Mwaka V. M. (1990) *The Agricultural Utility of Wetlands of the Lake Kyoga Basin* in *Eastern and Southern African Geographical Journal*, Vol. 1 No. 1; Zie Gariyo (1991), *Appropriate Technology, Productivity and Employment in Agriculture in Uganda: A Case Study of the Impact of Kibimba and Doho Rice Schemes, Centre for Basic Research,* Working Paper No. 13; Gariyo Zie *(1992), Peasants, Technology and Transformation of Agriculture in Uganda* unpublished seminar paper presented to the Department of Sociology, University of Lund; Asowa-Okwe (1993*), Women Wage Workers in Plantation Estates in Uganda*, an OSSREA unpublished research paper.
2. Esther Boserup (1970) *Women's Role in Economic Development*, London: Earthscan Publications Ltd.
3. Claire Robinson and I. Berger (186) *Women and Class in Africa*. New York: Africana Press.
4. Claire Freeman *Colonialism and the Formation of Gender Hierarchies in Kenya,* a Review Article in Critique of Anthropology Vol. 7 No. 3.
5. See Introduction by Jane Parpart and Kathleen Staudt in Jane L. Parpart and Kathleen A. Staudt (1990) (eds.) *Women and the State in Africa*, Boulder and London: Lynne Rienner Publishers.
6. Bryceson and Mbilinyi (1980) quoted in M. Lovett (1990) *Gender Relations, and the Colonial State in Africa* in Parpart and Staudt (1990) *op cit.*
7. Lovett M. (1990) *Gender Relations, Class Formation and the Colonial State in Africa* in Jane Parpart and Kathleen Staudt. *op. cit.*
8. Claire Robertson and Iris Berger eds. (1986) *Women and Class in Africa*, New York: Africana.
9. M. Mamdani (1990) *A Methodological Note on the Historical Dimension of Gender Research*, an unpublished seminar paper. The Centre for Basic Research Gender Study Group.
10. E. Nnabuguzi *Peasant Response to Economic Crisis in Uganda; Rice Farmers in Busoga* in Review of African Political Economy No. 56, 1993.
11. Nnabuguzi (1993), *op cit.*
12. Fieldwork interview. 10 bags is equivalent to 50 – 60 tins of paddy. Each tin of paddy is about Shs. 5000/= which gives the peasant Shs. 250,000 per season.
13. See similar arguments by Prem Chowdhry, 1992 *High Participation Low Evaluation: Women and Work in Green Revolution Province of India (Haryana)* Unpublished seminar paper, paper presented to a Workshop on Women and Work: Historical Trends, Kampala, September 7 – 10. See also, other arguments by Phiona Macphail and Paul Bowels (1989) in *Technical Change and Intra-Household Welfare: A Case Study of Irrigated Rice Production in South Sulawesi, Indonesia* in Journal of Development Studies Vol. 26 No.1.

14. Fieldwork interviews.
15. This issue has been raised by Mwaka, (1990); Nnabuguzi, (1993) and Gariyo (1992). It is now an acknowledged fact that bird scaring in rice cultivation has contributed to peasant children not going to school in Eastern Uganda.
16. Prem Chowdhry (1992) *op cit.*
17. Fieldwork Interviews.
18. These studies are part of the Land Project at Centre for Basic Research and are published in a series of Working Papers. These are Kafureka Lawyer (1992) *The Dynamics of the Land Question and its Impact on Agricultural Productivity in Mbarara District, Uganda*, Centre for Basic Research Working Paper No. 25; Mugisha Robert (1992) *Emergent Changes and Trends in Land Tenure and Land Use in Kabale and Kisoro Districts*, Centre for Working Research, Working Paper No. 26; Nyangabyaki Bazaara (1992) *Land Policy and the Evolving Forms of Land Tenure in Masindi District,* Uganda, Centre for Basic Research Working Paper No. 28.
19. See Zie Gariyo (1991) *op.cit.* In our study of the Doho and Kibimba rice irrigation schemes, it was noted that because women had access to modest incomes, their consciousness was beginning to change. They were beginning to challenge some of the traditional practices like polygamy, family labour of women. And because women were able to employ men as casual and contract labour on their own plots, women's control over their labour power was no longer taken for granted.
20. This information was given to me by N. Mwaka at a seminar to discuss Employment Policies at Mbale, Uganda in November 1992.
21. In order to qualify for a loan peasants had formed themselves into a group with a few women members and had obtained a loan from the local Uganda Commercial Bank in Iganga.
22. In Zie Gariyo *op. cit.* (1991), we were informed by the women rice cultivators at Kibimba of the stiff loan conditions they were facing and the heavy debt which they had accumulated because the bank had intentionally released the cash portion of the loan when it was too late, i.e. at the time of harvest instead of at the time of transplanting, weeding and bird scaring. Bank strategy, they were convinced, was merely out to get a profit.
23. During the course of fieldwork we came upon some of these clubs like Bakussekamaja Women's Club composed of 15 women members. They provide on the group farm every Wednesday and have been growing soya, cassava, Irish potatoes, and rice. Other such groups include one called Kiwanyi Twabiwulila Women's Association; Tugezeku Women's Club with 15 members which started in 1989; Kamukamu Women's Club with 15 members formed in 1992. A cursory look at the membership composition indicates that most of these women are ex-teachers, wives of local bureaucrats, O'level leavers. Very few clubs include completely non-literate women. Most illiterates prefer the less complicated labour groups we discussed earlier on.
24. This study is also part of a larger study, which we are undertaking within

peasant production systems, on the impact of irrigated agriculture on peasant production systems. Another study on its impact on peasant food systems will also be undertaken.

25. Some enterprising young men however, have improved their housing conditions by constructing brick and tinned roof houses, purchasing an extra piece of land, setting themselves up in commercial enterprises especially small trade in agricultural produce and shop keeping. Others earn incomes by using the bicycle for local transportation, which is the common means of hired transport in these parts of the country.

26. Joke Schrijvers (1988) *Blueprint for Undernourishment: The Mahaweli Development Scheme in Sri Lanka* in Bina Agarawal (ed.) *Structure of Partriarchy; The State, the Community and the Household*, London: Zed Books, 1988.

27. See for instance, Judith A. Carney (1988) *Struggles Over Crop Rights and Labour within Contract Farming Households in a Gambian Irrigated Rice Project* in Journal of Peasant Studies, Vol. 15 No. 3 1998. Gillian Hart *Engendering Everyday Resistance: Gender, Patronage and Politics in Rural Malaysia* in Journal of Peasant Studies, Vol. 19. No. 1 1991.

28. Findings from our earlier study on Kibimba and Doho rice schemes where rice production is semi-commercialised, still indicated that most hired labour is casual labour and therefore limited to specific agronomic practices.

29. During field work, only six households indicated that they hire seasonal labour for rice cultivation.

30. Fieldwork interviews.

31. See Gillian Hart (1991), *Engendering Everyday Resistance: Gender, Patronage and Production Politics in Rural Malaysia* in *Journal of Peasant Studies*, Vol. 19 No.1.

32. Peasants frown upon such women who have taken an independent action to set up their own households without necessarily being married to any or one man. Some of them raise children born from several men but refuse to marry any of them.

33 Resistance Councils (RCS) in Uganda are the country wide popularly elected local political cum security organs. They are elected from among the members of a locality with the village being the smallest unit at RCI. Women are specifically represented at the committee of nine people that is the executive of the village but they are free to contest the rest of the eight posts.

34. Field Work Interviews.

REFERENCES

A. Theses and Unpublished Material

Asowa-Okwe (1993), *Women Wage Workers in Plantation Estates in Uganda*, An OSSREA Unpublished Research Paper.

Asowa-Okwe (1993) *Irrigation Technology and Its Impact On Labour Dynamics in Doho and Kibimba Rice Schemes in Eastern Uganda.*

Unpublished Seminar Paper, Centre for Basic Research.

Gariyo Zie (1992), *Peasants, Technology and Transformation of Agriculture in Uganda.* Unpublished seminar paper presented to the Department of Sociology, University of Lund.

Chowdhry Prem (1992), *High Participation Low Evaluation: Women and Work in Green Revolution Province of India (Haryana)* Unpublished seminar paper, Paper presented to a Workshop on *Women and Work: Historical Trends*, Kampala, September 7 – 10.

Elson Diane (1991), *Gender Analysis and Economics in the Context of Africa* Unpublished Paper for CODESRIA Workshop on Gender Analysis and African Social Science, Dakar, September 1991.

Hamsworth J. (1962) *Dynamics of Kisoga Land Tenure.* Unpublished Conference Paper. East African Institute of Social Research, Kampala.

Mamdani M. (1984) *The Agrarian Question in Developing Countries.* Unpublished Paper presented to a Mawazo Workshop, Makerere University, Kampala.

Mamdani M. (1990) *A Methodological Note on the Historical Dimension of Gender Research.* Unpublished seminar paper to the Centre for Basic Research (CHB) Gender Study Group.

Tadria H. M. K (1985) *Changing Economic and Gender Patterns Among the Peasants of Ndejje and Seguku in Uganda*, University of Minnesota, Unpublished PHD Thesis.

Articles

Ahmad Zubeida (1980), *The Plight of Rural Women: Alternatives for Action* in Journal of International Labour Review, Vol. 119 No. 4.

Alaudin Mohamed and Clem Tisdell (1991), *Welfare Consequences of Green Revolution Technology: Changes in Bangladesh Production and Diet* in Development and Change, Vol. 22 No. 3.

Bhalla A. S. (1989) *Innovations and Small Producers in Developing Countries* in Economic and Political Weekly, February 25.

Bowen Merle L. (1989) *Peasant Agriculture in Mozambique: The Case of Chokwe, Gaza Province* in Canadian Journal of African Studies Vol. 23 No. 3.

Bujra, Janet (1983) *Class, Gender and Capitalist Transformation in Africa* in Africa Development. Vol. VIII No. 3 (July – Sept.) CODESRIA.

Byres T. J. (1977) *The New Technology, Class Formation and Class Action* in the Indian Countryside Journal of Peasant Studies Vol. 8 No. 4.

Brady Barbara, *Male Rationality in Economics – a Critique of a Godelier on Salt Money* in Critique of Anthropology, Vol. 3 No. 9 & 10.

Bryndon L. and Chant S. (1989) *Women in the Third World. Gender Issues in Rural and Urban Areas*, Edward Elgar, London.

Carney Judith (1988) *Struggles Over Crop Rights and Labour Within Contract Farming Households in a Gambian Rice Project* in Journal of Peasant Studies Vol. 15 No. 3.

____(1990) *Manufacturing Dissent: Work, Gender and The Politics of Meaning in a Peasant Society* in Africa Vol. 60 No. 2.

Deere, C. D. (1976) *Rural Womens' Subsistence Production in the Capitalist Periphery:* in *ROAPE*, Vol. 8 No. 1.

Dorthe Non Bulow and Anne Sorensen (1993), *Gender and Contract Farming: Tea Outgrower Schemes in Kenya* in Review of African Political Economy Vol. No. 56.

Edholm Felicity, and Olivia Harris and Kate Young (1977) *Conceptualising Women* in Critique of Anthropology, Vol. 3 No. 9 & 10.

Evans Alison (1989) *Women: Rural Development Gender Issues in Rural Household Economics*. Institute of Development Studies. Discussion Paper No. 254.

Hart Gillian (1991) *Engendering Everyday Resistance: Gender, Patronage and Politics in Rural Malaysia* in Journal of Peasant Studies, Vol. 19 No. 1.

Freeman Carla (199 ?), *Colonialism and The Formation of Gender Hierarchies in Kenya* in Critique in Anthropology, Vol. 7 No. 3.

Heald Suzette (1991), *Tobacco, Time, and the Household Economy in Two Kenyan Societies: The Teso and the Kuria* in Journal of Comparative Study of Society and History. Vol. 33 No. 1.

Kuhn, Annette and Wolpe, Annie Marie (1978) *Feminism and Materialism: Women and Modes of Production*. Routledge Keegan and Paul.

MacPhail P. and Bowles P. (1989) *Technical Change and Intra-Household Welfare: A Case Study of Irrigated Rice Production in South Sulawesi, Indonesia* in Journal of Development Studies Vol. 26 No. 1 October.

Mwaka V. M. (1990a) *The Diffusion of Irrigation Farming in Iganga and Kamuli Districts* in E.G.

____ (1990b) *The Agricultural Utility of Wetlands of the Lake Kyoga Basin* in *Eastern and Southern Africa* Geographical Journal Vol. 1 No. 1.

E. Nnabuguzi (1993) *Peasant Response to Economic Crisis in Uganda; Rice Farmers in* Busoga in Review of African Political Economy No. 56.

Naila Kabeer (1991) *Gender Dimension of Rural Poverty: Analysis from Bangladesh in Journal of Peasant Studies*, Vol. 18 No. 2.

Nzioki E. A. (1991) *Women in Food Production: Their Work and Constraints. A Case in Mumbuni (Machakos District – Kenya)* an OSSREA Occasional Paper No. 2.

Upadhayaya, H. K., *et. al* (1990) *Differential Adoption of Modern Rice Technology and Regional Wage Differential in Nepal*. In Journal of Development Studies, No. 3 April.

Quisumbing Agnes R. (1988), *Women and Agrarian 35 Transformation in the Philippines: Food Crops, Cash Crops and Technical Change*. Women's Resource and Research Center, Monograph Series No. 2.

Singh R. D. and Morey J. M. (1987) *The Value of Work - At Home and Contributions of Wives' Household Service in Polygamous Families: Evidence from an Africa LDC* in Economic Development and Cultural Change, Vol. 35 No. 4.

Secondary Sources

Abbot P. and Sapsford R. (1987) *Women in Social Class*, Tavistock Publications Limited, London, New York.

Agarwal Bina (ed.) (1988) *Structure of Patriarchy: State, Community and Household in* Modernizing Asia, Zed Press.

Bardhan P. (1982) *Land, Labour and Rural Poverty*, Colombia University Press.

Barrett M. (1988) revsd. edn. *Women's Oppression Today*, Verso.

Bhati U. N. (1976)–*Some Social Economic Aspects of the Introduction of New Variety Seeds in Malaysia* UNRSID.

Boserup E. (1970) *The Role of Women in Economic Development*, Earthscan Publications, London.

Boserup E. (1981) *Population and Technology*, University of Chicago.

Chimpungu N. Samuel (1988) *The State, Technology and Peasant Differentiation in Zambia: A Case Study of the Southern Province* 1930 – 1986. Historical Association of Zambia, Lusaka, Zambia.

Cutrufelli M. R. (1983) *Women of Africa: Roots of Oppression*, Zed Press.

Gariyo Zie (1991) *Appropriate Technology, Productivity and Employment in Uganda.* Centre for Basic Research. Working Paper No. 13.

E. G. Nnabuguzi, *The Agrarian Question and Technological Change in Uganda.* Makerere University, MISR.

Obbo Christine (1980) *African Women; Their Struggle for Economic Independence*, Zed Books.

Papart Jane and K. A. Staudt (1990) *Women and the State in Africa*, Lynne Reinner Publishers, Boulder and London, 1990.

Powesland (1957) *Economic Policy and Labour: A Study in Uganda's Economic History*, EAISR, Kampala, 1957.

Sacks Karen (1982) *Sisters and Wives: The Past and Future of Sexual Equality*, University of Illinois Press, Urbana and Chicago.

Robertson Claire and J. Berger (eds.), (1986) *Women and Class in Africa*. New York: Africana.

G. S. Were (ed.) *Women and Development in Africa*, G. S Were Press, Nairobi, Kenya.

B. Centre for Basic Research Working Papers

Kafureka Lawyer (1992) *The Dynamics of the Land Question and its Impact on Agricultural Productivity in Mbarara District, Uganda*, Centre for Basic Research Working Paper No. 25.

Mugisha Robert (1992) *Emergent Changes and Trends in Land Tenure and Land Use in Kabale and Kisoro Districts*, Centre for Working Research Working Paper No. 26.

Nyangabyaki Bazaara (1992) *Land Policy and the Evolving Forms of Land Tenure in Masindi District, Uganda*, Centre for Basic Research Working Paper No. 28.

5 Agricultural Land Management for Sustainable Production in Swaziland

Paul S. Maro

The Kingdom of Swaziland lies between latitudes 25 44 and 27 19 South and longitudes 30 47 and 32 07 East. The longest distance from north to south is 194 km. and from east to west is 145 km. Its area is 17,360 km sq. Table 5.1 shows the major characteristics of the four ecological zones and how relief, soils and rainfall affect the resource potential of these ecological zones. Apparently, there are two contradictory environmental management trends in Swaziland which, if not checked, can lead to a major crisis in resource use and affect drastically the sustainability of agricultural production. First, the size of the country is relatively small (17, 360 km sq.) and is inhabited by approximately 950,000 people. Land tenure is such that only 52% of the land is available to smallholder farms (Swazi Nation Land - SNL) inhabited by 70% of the population. It is estimated that 47% of the land is owned by 8 % of the populations with large-scale individual farms (ITF). Second, land use is ill planned or allocated to inappropriate uses coupled with the lack of land capability assessment measures to ensure long-term investments in land management.

Therefore, the core question this study set out to answer was, do the problems of land management generally observed in Swaziland. It is particularly interested in explaining land degradation, soil erosion, overstocking and their consequences. A point in focus is whether food production decline is a result of improper land use planning based on crop soil suitability. Conversely, is food production decline a result of usufruct/communal land tenure system that might have discouraged investments in land? Moreover, whether it is as a result of land? Is food production decline the result of low producer prices and low rural income that leads to off-farm employment?

Table 5. 2 shows the disproportionate increase in both the percentage of total population and density of population on Swazi Nation Land compared to Individual Tenure Land (ILT). A large proportion of SNL areas is located in the Middleveld (32% in 1976 as compared to 41% in Lowveld, 21% in Highveld and 6% in Lubombo). Obviously these areas have always carried the highest number of people (the percentage is declining) out of the four zones. The average density of population has nearly doubled between 1966 and 1986. Similar trends in change in density are observed for the Highveld and for Lubombo both of which have a comparatively more fragile land resource base because of the rugged terrain.

M.A.M. Salih (ed.), Local Environmental Change and Society in Africa, 2nd Edition, 101–128.

Table 5.1: *Characteristics of ecological zones*

Region	Area (sq.km)	Altitude (Metres)	Temper-ature (mean annual)	Rainfall (mean mm)	Vegetation
High-veld	5,029	1070-1460	17°C	1100-1850	Short sour grasses
		(1860)			Patches of mountain
		Granites			Forest
		Steep 18%– 50%			Trees and shrubs
		Slopes			Commercial forests
		Broken terrain			
Middle-veld	4,957	54C - 915	18°C	870-1000	Tall moist or dry
		Igneous & metamorphic			tree savana. Mixed.
		Broken terrain			Acacia Karoo. Smallholder agriculture
		Steep 15% slopes			Good for cattle. Large plantations
Low-veld	6,416	Average 350 Complex geology Gentle slopes 3% Undulating.	22°C	520-675	Bushveld-Savana, acacia and broad-leaved trees tall grasses. Smallholder agriculture Large citrus & sugar plantation
Lubomb o	1,321	Average 765	20°C	865	Similar to Middle-veld mixed bush
		Intermediate volcanic lavas			
		Steep escarpment from Lowveld			

They depend on family labour and low-level inputs hence produce generally low yields. Individual tenure farms in contrast are large, use modern inputs and produce for the market. Some small-scale irrigation is practised on SNL in the Highveld and Middleveld but the main irrigation areas are for sugar cane in the Lowveld.

Land use is greatly influenced by the physical environment and by population density. Only about 15% of the land in Swaziland is arable although less than this is actually planted with crops. A good proportion of this scarce cropland is left fallow and this represents potentially productive land if intensive methods of land use are adopted. The largest proportion of the land on both SNL and ITF is, however, devoted to grazing but some of it can be used for crops especially when livestock numbers have been reduced to match carrying capacity of land. It is also estimated that 20% of the land under ITF is unused but sometimes loaned to small holders for agriculture (C.S.O. 1981). Except for tobacco and cotton, SNL farmers concentrate on the production of food crops such as maize, beans, sorghum, bananas, groundnuts, rice, pineapples, vegetables and fruits.

Table 5.2: *Population distribution by land tenure and ecological zone*

Distribution by	% of Total Areas[(1)]	% of total Population			Population Density per km²)		
		1966	1976	1986	1966	1976	1986[(2)]
Land Tenure							
Rural (SNL)	52	-	66	70	28	36.1	-
Farm (ITL)	47	-	19	8	9	11.5	-
Urban	1	-	15	22	252	396.6	-
Ecological Zone							
Highveld	29	28	31	32	21	30.7	43.0
Middleveld	26	42	41	38	28.2	33.9	51.8
Lowveld	39	25	24	24	14.6	18.5	25.2
Lobombo	8	5	4	6	14.7	16.3	30.7

Source: 1) Central Statistical Office: 1966 and 1976 Swaziland Population Census;
2) 1986 Census provisional figures and Maps

Notes: 1) Based on 1976 Census figures for land area
2) Own calculations based on provisional Census figure and Maps

Under the Swazi Nation system farmers can be allocated land for crop production only, and the area allocated varies considerably depending on the ecological zone, population density and size of the household. In 1972 the majority of the farms in all the ecological regions were between one and five hectares. By 1983 the most common size of cropped area on SNL was down to 0.5 to –1.0 ha, but with the Rural Development Areas (resettlement schemes) having a higher proportion of the farms over 3 ha compared to non-RDA areas.

The increasing imbalance between population and resources is also illustrated with the consideration of the average amount of land available per person on SNL. It has been observed that there is a steady decrease in the amount of arable land available per rural person from an average of 0.5 ha - 1.0 ha in 1960 to 0.10 - 0.20 in 1986, which are very uneconomical sizes. Because not everyone in the rural areas cultivates land, actual farm sizes are better than these calculated average figures. The implications of the deteriorating relationship between numbers of people and land resources nonetheless call for urgent action in population control to bring down the rate of natural increase. Land use planning is important to assess suitability. Intensification of agriculture is important to enable the small *per capita* farm sizes to produce more for the increasing population. Equally important is the diversification of the economy into non-agricultural sectors to absorb the population and check the influx from rural to urban areas. It is also desirable to reassess the current land tenure system with a view to transferring some more land from ITF to SNL.

Livestock are important on SNL. Over 500,000 cattle are on SNL and only about 100,000 are on ITL, while in 1984 other types of livestock totalled 298,000 goats, 700,000 poultry birds, 30,000 sheep, 15,000 horses, mules and donkeys and 20,000 pigs (FAO, 1985). The average herd size of cattle in 1983 was 18.5 but this differed between SNL with 16 head per household and 48 head per farm on ITL. Whereas pastures have been improved and grazing is managed on ITL ranches, communal grazing is practised on poor natural pastures on SNL, resulting in degradation and serious soil erosion. The average stocking rate in Swaziland, including cattle, sheep, goats and equines is 2.5 ha. per livestock unit (L U = 350 kg) with SNL communal grazing areas overgrazed by 50% at 1.9 ha per LU and ITL stocked at 4.5 ha per LU. The stocking rates on SNL differ within the ecological zones with the Highveld with 1.94 ha/LU; Middleveld, 1.33 ha/LU and 2.10 ha/LU in the Lowveld. These stocking rate figures would be much higher if only grazing land on SNL was considered. Estimated carrying capacity per LU is 2.65 ha for both the Highveld and Middleveld, and 4.0 ha for the Lowveld (World Bank 1977). Besides providing meat, milk manure, transport and draught power in tilling the soil, cattle are also used as "lobola" (bride-wealth) and are a store of wealth that can provide prestige. They are a form of investment and security for the farmer and often provide access to credit.

The most significant impact of population pressure on land is degradation and soil erosion. Most of the arable land has slopes of 3-12% with relatively shallow and fragile soils that are subject to severe sheet and gully erosion under the intensive rainfall and farming practices leaving the soils unprotected for most of the year. But the introduction in the 1950s of mandatory grass contour stripping between fields, contour ploughing, and the prohibition of cultivation on or near river-banks has greatly reduced erosion on arable land.

Without application of manure or fertilisers however, the soils have become infertile with continuous use and this has affected yields adversely. Maize production in Swaziland has declined from 110,607 tonnes in 1978 to 82,760 tonnes in 1983 due to falling production in SNL small-farms, which declined by 36% compared to a 48% increase in large scale (ITL) farms in the same period. Maize imports therefore increased from 35,088 to 61,781 tonnes between 1978 and 1983, and this represents 43% of the annual national consumption of maize (Ministry of Agriculture, 1984). If the population continues to increase at the present rate the country will have to import three times the amount of maize presently imported by the year 2000, unless fertilisers and agro-chemicals are judiciously applied to increase output without harm to the environment.

One of the more visible impacts of population pressure on land is severe soil erosion due to overstocking on communal grazing in SNL. The stocking density is 1.9 ha/LU which is double the livestock carrying capacity of the land. This has led to pasture depletion, widespread loss of soil due to sheet and gully erosion, bush encroachment, drying up of springs, dam silting and obviously low animal productivity. The rate of erosion may be as high as 25-35 tonnes of soil loss per hectare per year as compared to a maximum of about three tonnes per hectare per annum. According to some estimates about 3/8 of the communal grazing land (308.000 ha) has soils with a life expectancy not exceeding 50 years because the stocking rate is less than 2 ha/LU, and the rest of the SNL grazing land has a life expectancy of up to 100 years, that is if the stocking rates and soil erosion are not stopped and reversed. (Roder, 1977).

Table 5.3: *Pressure of population and arable land in SNL ecological total*
A. SNL Ttl SNL Arbl SNL Arbl Lnd ha/pr

Zone	(ha)	Area (ha)	Area (ha)	1960	1971	1976	1986 (2)
Highveld	502,950	200,362	19,195	0.72	0.27	0.20	0.12
Middleveld	459,750	318,234	45,284	1.11	0.34	0.34	0.21
Lowveld	641.620	352,603	38,085	1.32	0.55	0.41	0.3
Lubombo	132.120	60,295	3,833	0.45	0.24	0.21	0.10
Swasiland	1,736,440	931,494	108,397	1.00	0.36	0.31	0.20

- 1) Includes cropland, fallow and homestead area on SNL, and considering only rural population
- 2) Based own estimates of rural population from 1986 provisional census figures.

I have used two complementary approaches or models to assess land suitability and the human carrying capacity of land (critical density) which assesses the number of people per given unit of land, and therefore the amount of land required per household to achieve the desired objectives. It is not enough to know that a unit of land is suitable for maize production or for grazing. In order to maintain that land in sustainable production it is necessary to know the number of people or households it can support under specific conditions (or assumptions). The agro-ecological zones approach entails the following concepts and variables:

1) Climatic Resources Inventory including temperature and moisture (rainfall). Potential evaporation and water-balance diagrams can show the length of growing period (LGP) and therefore indicate the types of crops suited to the LGP.
2) Soil Resource Inventory for the sample areas was based on the existing 1:125,000 soil map with its local classification of "sets" and "series" which are adequately described to identify the crops suited to such soil units.

The combination of climatic and soil information will produce optimal land suitability for specific crops. The FAO-AEZ study has only four classes under anticipated crop yields: very suitable, suitable, marginally suitable, and not suitable for crops. In the micro application to the sample areas this approach helps to identify land use types that "match" and those that "do not match" the suitabilities identified. It is possible that land tenure arrangements in combination with socio-economic factors of family labour and prices paid to farmers for crop sales might explain the practice of putting land to uses to which it is not best suited.

The second and complementary approach aims at establishing the maximum number of people a given area can sustain permanently at a given standard of living without deterioration in the land resource.

The concept of critical population density, or population carrying capacity of land (Allen, 1965; Brookfield, 1972; Geertz, 1963; Maro, 1975; Newman and Matzke, 1984) is very useful in showing the "over" or "under-population" when the calculated carrying capacity is compared to the actual population census, the number of people, but also the required amount of land to meet certain set goals. This population carrying capacity model uses three sets of variables:

1) Food requirements: the number of calories required per person per annum; the proportion of such requirements contributed by each crop grown; yields per hectare of crops grown; caloric value of crops per unit weight of food prepared for eating. If, for example, the caloric requirement per person is fixed at 2000 calories per day and an average family (father, mother and children) has seven persons, the annual food requirement would be (2000 x 7 x 360) 5,110,000 calories per household.
2) Minimum monetary income per household, derived from yield per hectare, price, and proportion of monetary income each marketed crop contributes.
3) The area per household for cash crops, food crops and grazing needed to achieve the caloric and monetary requirements set above.
4) Other assumptions would have to be made concerning how to classify and count areas that practice some form of mixed crop and livestock rearing on the same piece of land. It would be reasonable to count them twice. It is also assumed that the cattle are for meat rather than milk production.

The carrying capacity model (see Maro 1975: 19-21, 43-45) is suitable for small area studies for which most of the data is available. But for large areas such as whole big countries massive data and many calculations would be required. The great advantage of the model is that it is dynamic in that it accounts for changes in technology, prices, as well as in yields and introduction of new crops. It can be applied to different time periods for the same area, and can be replicated in any area provided data are available. The only setback with the model is that it depends on the assumptions and constants set, hence these have to be finalised only when some pre-test or cross checking has been done to ensure that they are realistic to the areas under study.

Both models reflect ideal conditions, which are rarely obtained in real life. Indeed in the present study it was possible only to work out approximate crop suitability according to the FAO-ZEZ method, while scarcity of accurate and reliable data made it impossible to calculate the population carrying capacity of land according to the model outline above.

However, data obtained from the field interviews gives useful information on both farmer-perceived land suitability for crops and on perceived carrying capacity and land adequacy. This farmer-perception is an important variable in attempts to introduce better land management practices. Three areas were sampled in different ecological zones, namely, Gege in the Southwestern Highveld in Shiselweni District; Zombodze in the Middleveld in Manzini District; and Maphilenga in the Lowveld in Lubombo District.

Zombodze is situated almost at the centre of Swaziland, a few kilometres northwest of Manzini city. On the west the area is bordered by the Mdzimba mountain range. Within its area of about 82 km^2, the average altitude is 700 metres above sea level. The area experiences hot wet summers (16 – 26° C) and cool dry winters. Average annual rainfall is 900mm, which falls in summer between October and March.

The area is generally low-lying and undulating and is drained by several streams, the largest being Mtlane and Usushwana rivers. Natural vegetation is scattered mainly along the rivers and the rest of the area has man-made and some exotic vegetation, due to dense human settlement and intensive agricultural land use.

The average population density is 200 persons per square kilometre. Most of the people in Zombodze are in- migrants from other parts of the country who have settled permanently or temporarily in the area while the majority of them find employment in Manzini city and Matsapha industrial areas.

Zombodze is under Swazi Nation Land in the Central Rural Development Area, which was among the first RDA's to be established in the Kingdom. The RDA has had some impact on agriculture by providing agricultural machinery, extension and advice on agricultural practices, provision of farm inputs and assistance to farmers in selling their agricultural produce.

The area is well provided with health and educational facilities, clean water and good secondary roads, and is increasingly receiving more urban influence. A variety of crops are grown but maize and vegetables are the dominant crops. Vegetables are sometimes grown under irrigation. Cattle are important but grazing land is becoming increasingly scarce.

Gege area borders on the Republic of South Africa in the west, the Mbabane-Nhlangano road in the east and Geogea forest plantation in the south. Seventy percent of the area is Swazi Nation Land (SNL) and the rest is under private title deed ownership. The total area studied is about 40 square kilometres.

Gege lies between 100 and 1250 metres in altitude in the Highveld region of Swaziland. The climate is the humid semi-temperate type with temperatures ranging from -5°C in winter to 23° C in summer. The mean annual rainfall ranges between 1100mm and 2200 mm and occurs in the summer especially from October to February. Winters are relatively dry.

Unlike Zombodze, which is a receiving area for in-migrants, Gege is characterised by both out-migration particularly of young able-bodied males between 20 and 40 years of age to urban areas, and in-migration from surrounding areas and from South Africa. Maize is the main subsistence and cash crop grown, but beans, sorghum and potatoes are also grown. Livestock, especially cattle, goats and sheep, are widely raised in the area.

The population density in Gege is high between 150 and 200 per square kilometre, so natural vegetation is practically non-existent; some of the highland plains are covered with grasses and form part of the communal grazing grounds.

The area is provided with schools, health facilities and tap water, and is well connected to the major urban centres in Swaziland. *Maphilenga* area lies to the South of Sipophaneni off the main road to Big Bend. It is generally a low-lying area less than 500 metres above sea level. It experiences hot summers with a mean annual rainfall of 760 mm. Much of the natural vegetation of bushveld sawana, acacia and broadleaved trees and tall grasses is still found in most of the area.

Smallholder agriculture with cotton and maize as the main crops is expanding as people from other parts of the country continue to settle in the area. Cattle grazing is an important occupation. Unlike Zombodze and Gege areas, Maphilenga area does not seem to suffer from population pressure as settlements are few and dispersed and there seems to be plenty of unused land.

Social services, schools, health and water facilities are very dispersed and people have to cover greater distances than in the other two sample areas to have access to these services.

In addition to the already noted contrasts in the three sample areas in terms of altitude, rainfall, temperature and settlements, the three areas also differ in terms of soils and agricultural potential. Whereas in Zombodze the soils are generally deep, well developed grey and red sandy and clay loams (regosols, lithosols and ferralitic soil) and generally acidic; in Gege the soils are similar but in some places tend to be shallower than in Zombodze. Slopes also tend to be steeper and degradation and erosion of soil is more evident in Gege than in Zombodze. In Maphilenga the soils tend to be dominated by grey sandy loams which are shallow and only slightly acidic and some rock outcrops are more common than in both Zombodze and Gege.

DEMOGRAPHIC CHARACTERISTICS OF THE RESPONDENTS

Most of the respondents in all three villages were over 20 years of age, with the majority between 40 and 59. The implication of this age distribution is

Table 5. 4: *Age of respondents*

Age	Gege No.	%	Maph ilenga No.	%	Zomb odze No.	%	Total No.	%
20-29	4	7.4	0	0	0	0	4	2.6
30-39	8	14.8	7	18.9	6	10.2	21	14.0
40-49	11	20.4	10	27.0	15	26.0	36	24.0
50-59	17	31.5	5	13.5	20	34.4	42	28.0
60 +	11	20.4	6	16.2	11	19.0	28	18.7
ND	3	5.5	9	24.3	6	10.2	20	13.0
Total	54	100	37	100	58	100	149	100

that there should be no labour shortage for agricultural activities. The proportion of people 60 years and over was significant; 20% in Gege, 19% in Zombodze, and 16% in Maphilenga.

The number of respondents with no formal education was high in all the three villages, 51%, 39% and 24% respectively for Maphilenga, Gege and Zombodze. The majority of the respondents, except in Maphilenga, has primary school level of education, 46% in Gege, 45% in Zombodze, and 38% in Maphilenga. Therefore 89% of the respondents in Maphilenga, 85% in Gege and 69% in Zombodze had only up to primary school level of education. Post-primary school education was limited to 28% in Zombodze (1.7% had post primary school education), 11% in Gege and 10.8% in Maphilenga. Zombodze's proximity to both Manzini and Mbabane towns, which have a concentration of education facilities, and urban influence seemed to be an advantage, compared to Maphilenga. The level of education is supposed to influence agricultural practices.

CAREER PATTERNS

The majority of the respondents were farmers: 53% in Gege, 40.5% in Maphilenga and 43% in Zombodze. The next important occupations were labourer (20.3% in Gege and 17.2% in Zombodze), and miner (18.9% in Map- hilenga and 12.9% in Gege). Traditional healing and prophesying was important as an occupation in Zombodze only (6.8%) while 13.5% of the respondents in Maphilenga considered being a housewife as an occupation (table 5.5).

Table 5.5: *Main occupations of respondents*

Occupation	Gege No.	%	Maphi-lenga No.	%	Zombodze No.	%	Total No.	%
Farmer	29	53.9	15	40.5	25	43.0	69	46.3
Miner	7	12.9	7	18.8	2	3.4	16	10.7
Labourer	11	20.3	1	2.7	10	17.2	22	14.7
Housewife	2	3.7	5	13.5	3	5.1	10	6.7
Trad. Healer	0	0	1	2.7	4	6.8	5	3.3
Chief	0	0	2	5.4	0	0	2	1.3
Handicrafter	0	0	1	2.7	2	3.4	3	2.0
Driver	3	5.5	0	0	0	0	3	2.0
Soldier	1	1.8	0	0	1	1.7	2	1.3
Mechanic	0	0	1	2.7	3	5.1	4	2.6
Manager	0	0	0	0	1	1.7	1	0.6
Retired	1	1.8	0	0	2	3.4	3	2.6
Carpenter	0	0	1	2.7	0	0	1	0.6
Cook	0	0	0	0	1	1.7	1	0.6
Self Employed	0	0	1	2.7	1	1.7	2	1.6
Businessmen	0	0	1	2.7	1	1.7	2	1.3
Supervisor	0	0	0	0	1	1.7	1	0.6
Total	54	100	37	100	58	100	149	100

It is interesting to note that both Zombodze and Maphilenga have a higher proportion of their people gainfully employed compared to Gege which is further away from Manzini and Mbabane towns.

Over 70% of the respondents were married and only 15% were not married, and both the rates of widowed and divorced were very low (Table 5.6). Almost all the respondents had one wife, a few had two wives, and very few had more than two wives. This is contrary to popular beliefs that most Swazi men are polygamists: 91% in Gege had 1 or 2 wives, 100% in Maphilenga and 95% in Zombodze. In Gege 5.4% had more than two wives and in Zombodze there were 3.4% with more than two wives. 3.7% in Gege and 1.7% in Zombodze had no wives.

Table 5.6: *Marital status*

Status	Gege No.	%	Maphil-enga No.	%	Zombodze No.	%	Total No.	%
Married	42	77.8	27	72.9	48	82.8	117	78.5
Not married	8	14.8	7	18.9	2	3.4	17	11.5
Divorced	1	1.8	0	0	2	3.4	3	2.0
Widowed	2	3.7	2	5.4	0	0	4	2.7
ND	1	1.8	1	2.7	6	10.3	8	5.3
Total	54	100	37	100	58	100	149	100

The proportion of respondents with a small number of children, 1 to 2, born per family was small, 11% in Gege, 8% in Maphilenga and 10% in Zombodze. Over 40% in Gege and in Zombodze had 3 to 6 children born per family while 68% in Maphilenga were in this modal category. A significant proportion of the respondents, 30% in Gege, 36% in Zombodze and 13% in Maphilenga had 7 to 10 children born per family

The number of respondents with over 10 children born per family was 16% in Gege, 11% in Maphilenga and 9% in Zombodze. These respondents indicated a high birth rate and large family sizes in the three sample areas.

It was necessary to find out the proportions of the children born per family that were still alive as the majority of the children born per family actually were. Generally, according to these responses, about 3% of the children born per family died, and this is a fairly high infant and child mortality rate.

In order to assess the size of the household and ultimately the demands that are likely to be made on the land in terms of food and other needs, it was necessary to find out the proportion of children born and alive per family who were actually living with their parents. About 15% of the respondents in Gege and in Zombodze had no children living with their parents, and only 5% in Maphilenga. This means that the majority of the children in the three sample areas lived with their parents; respondents with 1 to 2 children living with their parents represented 39% in Gege, 22% in Maphilenga and 24% in Zombodze. Families with 3 to 4 children living with their parents were 28% in Gege, 38% in Maphilenga and 14% in Zombodze. The proportions with 5 to 6 children living with their parents were 15% in Gege, 11% in Maphilenga and 19% in Zombodze.

Because the majority of the respondents were over 40 years old (Table 5.4), it would be expected that some of their children should be employed and possibly contributing to the up-keep of the family. The responses showed that 41% in Gege, 51% in Maphilenga and 38% in Zombodze had no children who were employed. But 33% in Maphilenga, and 31% in Zombodze had 1 to 2 children who were employed; and the proportions with 3 to 4 children employed were 20% in Gege, 11% in Maphilenga and 9% in Zombodze.

While Gege and Maphilenga had 2% respondents with 5 to 6 children employed, Zombodze had 10% in this category, and an additional 3% with over 6 children employed, further illustrating its locational advantage of proximity to the two main urban centres of Mbabane and Manzini.

When numbers of the extended family (close relatives) living permanently with the nuclear family are added to the family size, we get the size of the household. The sizes of the household as reported by respondents show that a small proportion (less than 7%) of household sizes are between 1 and 4 members. The majority, 54% in Gege, 68% in Maphilenga, and 43% in Zombodze had households with 5 to 6 members. Proportions of household sizes with 11 to 16 members were significant, 26%, 22% and 34% for Gege, Maphilenga and Zombodze respectively. Large household sizes with over 16 members per household were 11% in Gege, 5% in Maphilenga and 12% in Zombodze.

The implications of large households for land use and satisfaction of basic needs, cast doubts on the ability of the household to satisfy their needs entirely from agricultural activities. Hence there was the need for wage employment for significant proportions of members of households in the three sample areas.

FARMLAND AND LIVESTOCK OWNERSHIP

The respondents were asked to state whether they possessed a farm. 94% in Gege, 86.4% in Maphilenga and 93.1% in Zombodze answered that they had a farm; the rest did not have farms. The farms, however, were small, 63% in Gege, 30% in Maphilenga and 48% in Zombodze had farms of less that 2ha. A significant proportion had farms that varied between 2 and 4 hectares 30%, 35% and 26% in Gege, Maphilenga and Zombodze respectively.

A small proportion had farms of 4 to 6 hectares. 7.4% in Gege, 8% in Maphilenga and 14% in Zombodze. Farm sizes over 6 hectares were found in Maphilenga (10.8%) and in Zombodze (8.5%) only. The rest did not respond. These sizes of farms do not include the communal grazing lands.

Asked whether they had cattle or not, the majority of the respondents, 31% in Gege, 35% in Maphilenga and 45% in Zombodze chose not to answer the question. Reluctance to reveal real numbers of heads of cattle is generally based on fears of increased taxation or reforms connected with the RDA programme. Of those who responded to the question, 18.5% in Gege, 8.1% in Maphilenga, and 13.8% said that they had no cattle. Those who had between 1 and 5 head of cattle were 13% in Gege, 19% in Maphilenga and 8.6% in Zombodze. A significant proportion, 26%, 11% and 16% respectively for Gege, Maphilenga and Zombodze had 6 to 11 head of cattle, while only 11%, 16% and 8.6% for the three areas had between 12 and 20 head of cattle. One respondent in Zombodze had 30 head of cattle and two in Maphilenga had 35 head of cattle. Cattle are important in Swazi society both in the socio-cultural ("lobola" gifts, during celebrations, esteem) and the economic (can be sold for cash, provide meat and milk, provide draught power and manure for agriculture) sense.

Judging from direct field observations and unstructured conversation with respondents, it is reasonable to conclude that there were many more cattle per household than reported in Maphilenga and in Zombodze, but Gege cattle numbers were low, partly because of land shortage.

MAJOR SOURCES OF INCOME

The major sources of income were reported as wage employment, sale of farm produce and a combination of the two. Beer brewing, handicrafts and healing were of secondary importance. Gege lies on the border with the Republic of South Africa and most of the respondents (38.9%) find jobs in both the Republic and in the public and small private sectors of this small town. Only 20.3% derived income from farming activities in Gege. This provides some interesting contrast to Maphilenga and Zombodze where income from farming activities was 32.4% and 32.8% respectively. A significant number of respondents in Maphilenga (32.4%)and in Zombodze (22.4%) derived their income from wage employment. Small proportions of the respondents, 16.7% in Gege, 10.8% in Maphilenga and 17.2% in Zombodze had income from both farming and wage employment. Clearly wage employment as a source of income was most important in Gege compared to income from farming activities, while farming activities were a more important sources of income for Zombodze compared to income from wage employment. For Maphilenga farming activities and wage employment were equally important as sources of income. It is significant that cattle sales were not reported as a regular source of income (table 5.7).

The implications of these findings are that farming activities cannot satisfy the income needs of the respondents, and unless steps are taken to increase the productivity of land and pay farmers an economic price for their produce, the tendency to opt for wage employment will continue to the detriment of agriculture production.

LAND TENURE AND MANAGEMENT

The dominant soil types in Zombodze, Gege and Maphilenga sample areas are deep pale grey sand on clay (regosolic), deep dark brown clay (ferralitic), and deep red loam (ferrisolic and ferralitic) soils. Significant areas are, however, covered with lithosolic red loams, grey sandy loams and some with rock outcrops, all of which soils tend to be shallow. Like the deep ferralitic soils, these lithosols are also acidic to very acidic. A few other areas have marsh (hydromorphic) sand and clay soils that are acidic.

Table 5.7: *Major Sources of Income*

Source	Gege No	%	Maphil-enga No	%	Zomb-odze No	%	Total No	%
Farming	11	20.3	12	32.4	19	32.8	42	28.2
Wage Empl.	21	38.9	12	32.4	13	22.4	46	30.9
Farm and Empl.	9	16.9	4	10.8	10	17.2	23	15.4
Handi-crafts	4	7.4	2	5.4	4	6.9	10	6.7
Farm and HDCFS	1	1.8	3	8.1	3	5.2	7	4.7
Beer Brew	1	1.8	1	2.7	1	1.7	3	2.0
Farm, Brew	1	1.8	1	2.7	0	0	2	1.3
Healing	2	3.7	0	0	2	3.4	4	2.7
Farm and Healing	0	0	0	0	1	1.7	1	0.7
Handcraft and Empl.	0	0	0	0	3	5.2	3	2.0
Cattlesales	1	1.8	0	0	0	0	1	0.7
Farm and Mechanic	0	0	1	2.7	0	0	1	0.7
ND	3	5.6	1	2.7	2	3.4	6	4.0
Total	54	99.8	37	99.9	58	99.9	149	100

In Gege the dominant soils types include deep red and yellow loams (ferralitic) which are very acidic, grey sandy loam and marsh soil (lithosolos and hydromorophic); and rock outcrops and stony ground. These soils seem to present more constraints to their utilization compared to soils in Zombodze. Soils in

Maphilenga (figure 7) are more varied but seem to be dominated by four soil types; the deep pale grey sand on clay, deep pale red sand, shallow grey sand to sandy loam on hard rock, and the grey sandy loam on hard iron pan or on mottled clay pan.

The land suitability for crops and grazing has been summarised in Figures 8, 9 and 10. Most of the sample areas in Zombodze are categorized for crops and only small portions of the area are classified as "very poor soils on suitable slopes" (S2), and "unfit soils on slopes less than 14%" (S2). A very small area has been classified as "unfit soil on slopes more than 14%" (S3).

In summary, all of the sample areas in Zombodze are suitable for crops and for grazing, as areas classified as S2 are suitable for some crops and for grazing, while those classified as S3 may not be suitable for any crops but are suitable for grazing.

Compared to Zombodze, Gege area has a limited area with good soils on suitable slopes (S1); and significant proportions of the area have poor and very poor soils on suitable slope (S2) A limited area has unfit soils on slopes of more than 14%. Most of the constraints in Gege are steep slopes, and poor shallow soils. Although not suitable for crops, such land is suitable for grazing and forest plantations.

Most of the land in Maphilenga has very poor soil on suitable slopes (S2), with scattered large pockets of unfit soil on slopes of less than 14% (S2), very poor soil on suitable slopes (S2), and very limited areas with unfit soil on slopes of more than 14% (S3).

From the above description and maps, it is clear that most of the land in Zombodze is suitable for cultivation, while Gege's land is divided between suitable for cultivation or grazing. Maphilenga has very little land suitable for cultivation, while most of its land is suitable for grazing only. Through their accumulated experience, farmers in the three areas have learnt to identify the types of soils suited to specific land use types.

A large proportion of respondents, 43% in Gege, 39% in Maphilenga, and 33% in Zombodze either did not know how to identify soils for crops or did not respond. However, 26% of the respondents in Gege, 19% in Maphilenga and 28% in Zombodze said that soil colour is used to identify good soils for crops. Dark coloured soils were thought to be good soils by 14% in Zombodze, 5% in Maphielnga and 4% in Gege. Another 8% in Maphilenga and 7% in Zombodze said that they used soil colour and texture to identify good soils.

In Gege 7% used colour and productivity and another 6% colour and vegetation cover to identify good soils. Small proportions of respondents said that the texture, loam soils, humus and fertility were used to identify good soils. The farmers have accumulated knowledge from experience in associating the above soil characteristics with specific crops.

For example, when they were asked about soil suitability for maize production they responded as follows: 52% in Gege, 41% in Maphilenga and 47% in Zombodze said that red "sibovu" soils were best for maize production. Another 11%, 14% and 24% in the three villages respectively said that black loam soils are best for maize. Hence farmers perceived soil suitability for maize as red and black loam ("sibovu") soils. Large proportions, 35% in Gege, 27% in Maphilenga and 14% in Zombodze did not know how to identify soils suitable for maize or did not respond.

A similar question asked about how they identify soils suitable for cotton did not meet with good response because cotton is a significant crop only in Maphilenga. Hence 24% in Maphilenga said that "sibovu" and loam soils were suitable for cotton, while 94% in Gege 60% in Maphilenga and 95% in Zombodze said that they did not know, or had no response.

The farmers were asked to comment on trends in soil fertility, and 76% in Gege, 76% in Maphilenga and 64% in Zombodze said that soil fertility has declined. It is interesting to see their perception of what has brought this decline about and how they know that fertility has declined. 17% in Gege, 19% in Maphilenga and 5% in Zombodze said that soil fertility had declined due to lack of rainfall, while 4%, 3% and 5% in the three villages respectively said the decline in soil fertility was due to leaching.

Another 20%, 14% and 7% respectively attributed the decline in fertility to continued use of the soil, and 20% in Gege, 8% in Maphilenga and 19% in Zombodze said they had to use manure or fertilizer to stop the declining soil fertility. Significant proportions 15% in Gege, 11% in Maphilenga and 26% in Zombodze thought that there was no serious decline in soil fertility, it was more or less stable.

Asked about measures to correct the declining soil fertility, 56% in Gege, 41% in Maphilenga and 47% in Zombodze said they applied manure or fertilizers; 15%, 27% and 36% respectively said they did not do anything to correct the decline; and 28%, 30% and 14% respectively did not know or had no response to the question. Two respondents in Zombodze and one in Gege sought to correct declining soil fertility through better land management practices including control of soil erosion.

Agricultural production in the three sample areas is generally dependent on rainfall. When asked whether the rainfall in the last agricultural season (1988/89) was adequate, 83% in Gege, 76% in Maphilenga and 79% in Zombodze said it was adequate. Only 13%, 16% and 17% respectively said the rainfall was not adequate. The rest did not know.

The interpretation of adequate rainfall, according to the farmers, was a good or adequate harvest (39% Gege, 27% Maphilenga and 33% in Zombodze) and a well distributed rainfall (39% in Gege, 38% in Maphilenga and 3% in Zombodze). Only 8% in Maphilenga??, 3% in Gege, 24% in Maphilenga and 34% in Zombodze did not respond.

Asked whether there had been drought in recent years, 91% in Gege, 78% in Maphilenga and 84% in Zombodze said yes, there has been drought, and 4%, 11% and 12% said there had been no drought, while the rest did not respond.

The majority of the respondents perceived as drought the 1986 to 1987 period, 25% in Gege, 49% in Maphilenga and 69% in Zombodze. Another 20% in Gege had experienced drought in the 1983 to 1985 period, while an additional 22% in Maphilenga felt that they had experienced drought from the early 1980s. Only 4% in Gege and 3% in Zombodze said that they experienced drought every year. The rest said they did not know. It is clear that only Zombodze seems to have experienced drought only recently while Gege in the south west experienced drought from 1983 to 1989 and Maphilenga in the lowveld seems to experience some drought every year.

But when asked about the duration of the perceived drought a significant proportion (32% in Gege, 32% in Maphilenga and 21% in Zombodze) said that it lasted less than "a few months", in any case less than one year. Another group said that the drought lasted between one and three years (24%, 11% and 31% respectively), while another 11% in Gege, 5% in Maphilenga and 22% in Zombodze said it lasted "a whole crop season". The rest (31%, 51% and 28% respectively) did not know or did not respond.

The farmers in Zombodze and Gege seem to have experienced real drought conditions for three years, while those in Maphilenga seemed to refer to their characteristically low rainfall pattern in which sometimes the first rains often come later than expected. None of the farmers, however, perceive rainfall and drought as a limiting factor in agricultural production.

LAND USE AND MANAGEMENT

According to the responses to the interviews, land in Swaziland is acquired primarily from the chief acting on behalf of the king. In Gege 66% of the respondents obtained land from the chief gratis, while 26% obtained land from the chief after paying some tribute, usually a cow. In Maphilenga 48% had obtained land gratis from the chief and 35% had paid some tribute in order to get land. By contrast, only 28% had been given land freely by the chief in Zombodze while 52% had to pay some tribute to the chief to get land. This might reflect not only increasing pressure of population and land scarcity but also possible commercialisation of land in Zombodze because of the urban influence. The only other reported methods of acquiring land were inheritance (4% Gege, 14% Maphilenga and 16% Zombodze), and direct grant from the king or the government (4% Gege, 3% Maphilenga and 2% Zombodze). Surprisingly, no one reported having acquired land by buying it, as this is against the law. But some tribute paying might amount to indirect buying of land.

Once land has been acquired, the owner has the right to build permanent housing, cultivate crops, lease the land for short periods, and pass it on to his children through inheritance.

All respondents in the three sample areas reported that these were the basic rights to land as recognized by the chief and the community. Although these are usufruct rights they are tantamount to full ownership of the land.

However, there are certain conditions which can lead to land being taken away or to someone being expelled from the chief down: witchcraft (50% Gege, 59% Maphilenga and 65% Zombodze); being a nuisance to the community (31% Gege, 16% Maphilenga and 16% Zombodze); contradicting the chief or showing disloyalty (11% Gege, 16% Maphilenga and 5% Zombodze). Resettlement was also mentioned as a condition whereby someone could be moved from the land in Maphilenga (8%) and in Zombodze (10%). This referred to the Rural Development Areas Programme (RDAP) where some people were moved from their land but were also allocated new areas.

The other conditions mentioned as being attached to land included the fact that someone had to work for the king and/or the chief when summoned to do so (2% in Gege, 19% in Maphilenga and 28% in Zombodze) and that someone had to pay tribute to the chief (39% in Gege, 8% in Maphilenga and 10% in Zombodze) on certain occasions. It is generally believed that the traditional usufruct tenure on cropland could be a hindrance to improvements on the farm (ref. Hughes, 1972; de Vletter, 1983; Armstrong, 1985). However, only a very small proportion of the respondents in Gege (4%) and Zombodze (3%) said that land laws, especially the inability to sell land, was a hindrance to improvement of land.#

The majority (35%, 38% and 43% in Gege, Maphilenga and Zombodze, respectively) cited lack of capital as the major constraint, while another 19% in Gege, 40% in Maphilenga and 10% in Zombodze saw climate and lack of water as another major hindrance to land improvement. Land scarcity and low level of output were also cited as hindrances to improvements in Gege (19%) and Zombodze (10%). A few respondents in Gege (6%) and in Zombodze (10%) felt that there were no hindrances to making improvements on land.

When asked what improvements and changes in land tenure they were suggesting or how they would like to promote good land use, the majority of the respondents (54% in Gege, 50% in Maphilenga and 53% in Zombodze) did not respond. But 11% in Gege, 5% in Maphilenga and 9% in Zombodze suggested that individual private ownership of land would encourage more productive use. Another 2% in Gege, 41% in Maphilenga and 33% in Zombodze said that they needed the services of extension officers. Only 33% in Gege, 3% in Maphilenga and 5% in Zombodze saw no need for change in land tenure.

Grazing land is communally owned, used and supposedly managed. While 40% in each sample area did not have any response to the question of a person's rights over grazing land, 56% in Gege, 51% in Maphilenga and 50% in Zombodze recognized that they had free and full rights to use the communal lands for grazing and for fuel wood. How and by whom are communal grazing lands managed? A significant proportion of the respondents reported that the communal grazing lands are not maintained by anyone (43% in Gege, 65% in Maphilenga and 33% in Zombodze). In Gege, range management is practised through rotational seasonal grazing (19%) made possible partly by fencing (6%). Only 3% in Maphilenga and 2% in Zombodze reported rotational grazing. Another 14% in Zombodze said that communal grazing areas were managed by the community and led by the chief. The "no response" rate for this question was high; 31%, 32% and 43% respectively for Gege, Maphilenga and Zombodze.

As a summary, when the respondents were asked whether the system of land acquisition, ownership and inheritance was conducive to good and improved land use, 91% in Gege, 92% in Maphilenga and 90% in Zombodze responded "yes", and only 6%, 8% and 7% respectively thought that the system was not conducive to good land management practices.

AGRICULTURAL PRODUCTION

The proportion of respondents who are native to the areas is significantly low (13% in Gege and 10% in Zombodze) and in Maphilenga none of the respondents was native to the area. Although most settlement took place in the period 1960's to 1980's in the three areas, it is evident that Zombodze has had settlements since the 1920's and Gege since the 1950's while in Maphilenga settlements are very recent, 51% took place in the 1980's.

Reasons given by respondents for preferring to settle in these areas were: liked it (33.3% Gege, 27% Maphilenga and 34.4% Zombodze); it was the place chosen by the chief (22% Gege, 11% Maphilenga and 19% Zombodze); inherited (6%, 11% and 12% respectively for the three areas); and through marriage (7%, 20% and 5% respectively). In Gege and Zombodze (5%) some said they were born in the area, and the rest (19%, 29% and 24% respectively) did not respond.

The majority of the respondents use the land for growing crops. 69%, 70% and 72% respectively use the land for growing crops while another 28%, 19% and 22% respectively use the land for other productive activities.

The main crops cultivated in the area include maize, cotton, pumpkins, beans and vegetables. These crops are grown because they are the staple (60%, 22% and

38% for the three areas respectively); and because there is no alternative to growing these crops (6%, 32% and 3% respectively). Maize is the dominant crop everywhere because it is the staple that is also sold as a cash crop. Cotton is important only in Maphilenga area that has suitable soils (fig. 10) and practically no population pressure. Some cotton is also grown in Zombodze area. Vegetables including beans are an important source of income in Zombodze because the demand from the two major urban centres keeps increasing. Gege area also produces vegetables some of which are also marketed across the border.

Minor crops mentioned include beans in all the three areas, groundnuts in Maphilenga, sweet and round (Irish) potatoes in Gege and Zombodze, sugar cane in Zombodze and Zombodze has dairy produce.

For how long have these farmers been growing these major and minor crops? The responses follow the pattern of settlement history; 28%, 35% and 48% in Gege, Maphilenga and Zombodze have "always" grown these crops, or have grown them "for a long time", whereas 46%, 51% and 31% respectively started growing these crops in the 1970's and 1980's. Only 12%, 3% and 10% respectively started growing the crops between the 1940's and 1970's. With such a short settlement history there have not been any changes in the crops grown in the three areas. Most of the crops are grown singly in each field: 50% in Gege, 62% in Maphilenga and 72% in Zombodze.

37% of the respondents in Gege, 22% in Maphilenga and 19% in Zombodze practises inter-cropping, particularly of maize and beans, maize and pumpkins, maize and cotton and maize and potatoes. Where farmers own several fields in different locations, they might plant some crops singly in one field and interplant in other fields. Some crop rotation is practised, especially in Zombodze area.

The main agricultural inputs include seeds, manure, chemical fertilizers, machinery and labour. Seeds are easily available from the cooperative societies and from most shops in town, and their prices are generally known and affordable. As input, seeds are not discussed further in this report.

For the 1988/89 agricultural season, 33%, 22% and 29% of the respondents in Gege, Maphilenga and Zombodze each used from 10 to 59 bags of manure, and 65%, 70% and 59% respectively did not respond. The rest (2%, 8% and 12% respectively) each used between 60 and 300 bags of manure. However, the manure was applied on small plots, generally varying between 1 and 10 hectares in each case. The main crops on which manure was applied were maize (31%, 27% and 26% in the three areas), and cotton (14% in Maphilenga and none in the other two areas). Obviously manure was used largely by those farmers who own cattle; 16% in Maphilenga said the manure was free and 84% did not respond; 4% in Gege said the manure cost them E40; while 9% in Zombodze said it was free and 7% said it cost them E40, and the rest did not respond.

The farmers who used manure were aware that the use of manure resulted in increased production (8% Gege, 16% Maphilenga and 12% Zombodze). Another 3% in Maphilenga and 5% in Zombodze did not see much effect of the manure.

Whereas 35%, 30% and 41% of the respondents in Gege, Maphilenga and Zombodze respectively used manure on maize and cotton mainly, chemical fertilizers were used by more farmers than those who used manure especially in Zombodze and in Gege. In Maphilenga 84% of the respondents did not use chemical fertilizers while only 33% in Gege and 55% in Zombodze did not use fertilizers. The majority of those who applied fertilizers used small quantities of chemical fertilizers, one to ?? bags (48%, 24% and 62% for Gege, Maphilenga and Zombodze) and only a few used up to 20 bags (30% Gege, 3% Maphilenga and 19% in Zombodze).

Fertilizers were applied to maize (6%, 3% and 36% respectively for the three areas), maize and cotton (5% in Maphilenga only), maize and beans (2% in Zombodze only), beans (3% in Zombodze only) and cotton (5% in Maphilenga only). Only 2% in each of the three areas applied chemical fertilisers to "all the crops".

Chemical fertilizers are expensive when compared to animal manure. Of those who used chemical fertilizers, (66% in Gege) each spent up to E 250, while 3% and 24% spent that amount in Maphilenga and Zombodze respectively. Another 8% in Maphilenga and 7% in Zombodze spent up to E550 on fertilisers and another 3% spent up to E850 in Maphilenga. The effects of the application of fertilisers were: an increase in soil fertility and crop output (2% Gege, 15% Maphilenga and 33% in Zombodze); no change in output, 2%, 3% and 2% respectively; and no response (96%, 83% and 66% respectively).

Although not all farmers use fertilizers and manure, most did respond to the question of why they use these inputs: to increase crop output and/or yield (75% Gege, 32% Maphilenga and 75% Zombodze); to improve soil fertility (6%, 14% and 2% respectively); because they are cheap and abundant (9%, 5% and 10% respectively) referring to manure; lack of manure (6%, 5% and 3% respectively) referring to those who use fertilisers.

When asked about when they started to use these inputs, again the responses reflected the settlement history: 27%, 19% and 26% in Gege, Maphilenga and Zombodze respectively had "always" used manure and fertilizers, while the highest proportion, 44% in Gege, 40% in Maphilenga and 44% in Zombodze, responded that they started using them in the 1970's and 1980's. Another 7%, 3% and 9% respectively started to use these inputs in the 1950's and 1960's, and the rest did not respond.

Only 6% in Gege, 11% in Maphilenga and 8% in Zombodze have discontinued the use of these inputs mainly due to shortage of money and lack of cattle. Otherwise the rest of the respondents would like to continue to use these inputs to increase their crop yields.

Both manure and fertilizers are available within accessible distance from the farms. The C.C.U. and cooperative shops are the major sources of fertilizer (8.3% in Gege, 24% in Maphilenga and 78% in Zombodze), and homestead Kraal manure and manure collected from the farm supply all the manure used (11% in Gege, 30% in Maphilenga and 32% in Zombodze).

Asked about soil conservation measures they carry out on their farms, 33%, 27% and 24% in Gege, Maphilenga and Zombodze respectively responded that they do not practice any specific soil conservation; 54%, 22% and 48% respectively said they maintain grass strips usually planted on contour bunds; 7% in Gege, 27% in Maphilenga and 7% in Zombodze said they "add soil" to the farms, especially for vegetable gardens; and 4%, 11% and 16% respectively practise controlled furrow irrigation.

It is obvious that these conservation measures in combination with the application of manure and fertilizers by the majority of farmers in Gege (83% use manure or fertilizers) and in Zombodze (76%) and a few (24%) in Maphilenga ensure sustainability of the agricultural production systems.

The second type of agricultural input is the use of agricultural machinery for ploughing, planting and weeding. Swazi agriculture is fairly well mechanized. The most common machinery is the tractor with its different attachments for ploughing, harrowing/ridging, planting, weeding and transport. It costs from E50 at the beginning of the season to E100 at the peak, to plough one hectare of land. In the RDA's tractors are available for hire, but many farmers also own tractors which they also hire out. Most farmers plough from one to six hectares of land (57% Gege, 19% Maphilenga and 71% Zombodze), so they would have spent anything from E300 to E600 to plough their land.

Lack of oxen and no other alternative to machinery masks the fact that the Swazi farmer values his cattle to the extent that he is reluctant to use them for ploughing. Very few farmers use their cattle as oxen for ploughing and for transport as is now the case in many African countries. It is also true that tractors are more easily available and affordable to Swazi farmers than they are in many other African countries. Tractors, fertilisers and manure are becoming an integral part of the Swazi agricultural tradition.

LABOUR INPUT AND WAGES

The responses on labour inputs into different agricultural activities varied greatly among the three sample areas. The sources of labour for ploughing were the family for Gege (87%), the family and neighbours for Maphilenga (41% and 35%) and the tractor for Zombodze (55%). Family labour for ploughing contributed only 40% of the total labour in Zombodze, while only 2% and 14% in Gege and Maphilenga respectively used the tractor for ploughing.

Sources of labour for sowing seeds were: family (28%, 68% and 55% respectively); oxen (30%, 14% and 12% respectively); tractor (15%, 5% and 22% respectively); and neighbours (22%, 8% and 4% respectively). Weeding is, as can be expected, a more family labour-intensive activity and in Gege 54% used family labour for weeding. This might also be a reflection on the types of crops grown and the care with which weeding must be carried out. A small proportion, 15% in Gege and 8% in Maphilenga used neighbours, and another 17% in Gege, 3% in Maphilenga and 2% in Zombodze used a tractor for weeding.

Where crops have to be guarded against vermin it is family labour that was used. Similarly in harvesting the crop, family labour was used by 83% of the respondents in Gege, 84% in Maphilenga and 86% in Zombodze.

Hired labour for harvesting was used only by 9%, 5% and 2% of the respondents respectively. Transporting the harvest from the farm to the homestead was also done mainly by family labour, 56%, 65% and 71% respectively, while 5% and 9% in Maphilenga and Zombodze used hired vehicles to transport the crop home.

Only a few of the respondents hired labour for agricultural activities: 24% in Gege, 27% in Maphilenga and 22% in Zombodze. The majority (72%, 68% and 76% respectively) did not use any hired labour. Less than 15% of the respondents in each area hired up to five people, mainly for weeding (13%, 14% and 7% respectively) and clearing the land in preparation for ploughing (7%, 14% and 3% respectively). The cost of hiring labour was E15 per person per day, a price that is not prohibitive, as the labourer would work for less than a week to prepare or weed a field. This would amount to about E525 per season. But it should be noted that hiring labour is not as common as would be expected because the family labour supplemented by machinery is generally adequate for the farm size and activities involved. Indeed, when asked why they do not hire labour 50% in Gege, 51% in Maphilenga and 62% in Zombodze responded categorically that there was no need for hiring labour. But 17%, 5% and 21% respectively said that they did not have enough money to hire labour, implying an unfulfilled labour demand albeit a relatively low one.

To find out how else family labour is used, besides its use in agriculture, the farmers were asked whether they had any member of the family in wage employment. The majority of the families in the three areas 81% in Gege, 78% in Maphilenga and 76% in Zombodze had at least one member of the family employed outside the farm, and 7%, 22% and 21% respectively did not have any member of the family in wage employment. These responses do not conform entirely to the earlier findings where Gege had registered fewer people in wage employment compared to Maphilenga and Zombodze.

The member(s) of the family in wage employment were reported, and included fathers, mothers, sons and daughters, but in the majority of cases it was the father or son who was in wage employment. It is interesting to note that for Gege there are twice the number of fathers in wage employment compared to the number of sons while in Maphilenga there were more sons than fathers in wage employment, but in Zombodze the proportion was the same for fathers and sons. In Gege more daughters than mothers went for wage employment compared to Maphilenga where more mothers than daughters go for wage employment. In Zombodze the proportion of mothers to daughters in wage employment was almost the same.

Where do the respondents find off-farm employment? Not surprisingly, the majority find paid employment in Swaziland (14%, 40% and 59% for the three areas respectively) in the towns, rural public institutions and offices, and as labourers. Another 28% in Gege, 30% in Maphilenga and 3% Zombodze are employed in the Republic of South Africa; and 18% Gege, 5% Maphilenga and 2% Zombodze are employed in "various places" including a few soldiers in Gege. The rest did not respond.

The importance of off-farm employment was illustrated by responses to two questions. Do those employed supply some of their wages for family subsistence and for the purchase of agricultural inputs?

In Gege 76% of the respondents used their wages for family subsistence and agricultural activities, compared to 62% in Maphilenga and 69% in Zombodze. A not insignificant proportion, 13%, 14% and 21% respectively, said that they did not spend their wage earnings on subsistence and agricultural activities.

Scarcity of land and possible landlessness in Gege and Zombodze has made a large proportion of the male population in these areas seek wage employment to supplement earnings and production from agricultural activities. Similar conclusions were arrived at by previous studies for different areas of Swaziland (de Yletter, 1983; Russel, 1984).

The second question the responses of which illustrate the importance of off-farm employment was whether they would prefer wage employment to operating a farm and why. The majority, 63%, 59% and 50% respectively would not prefer wage employment to farming and 30%, 32% and 41% would prefer wage employment to farming. Note that a high proportion of respondents in Zombodze would prefer wage employment, partly because of insufficient land and their proximity to the major urban areas.

The reasons for preferring wage employment to farming and vice versa reveal that those who prefer employment do so because it brings in more money and in a less arduous way compared to farming. And this might be the reason why some would prefer employment when they are old, unusual as this might sound. On the other hand, the majority of the respondents in all the three areas prefer farming to wage employment because farming brings in more money, it is reliable, the whole family supports it and one works at home. The respondents would also encourage their children to operate a farm rather than go for wage employment.

The respondents were asked to make their own assessment of whether or not the available family labour was adequate for all agricultural activities. Maphilenga, which has a lot of unused land, had the lowest (54%) proportion responding that labour was adequate with a fairly high proportion (38%) saying family labour was not adequate for agricultural activities. In Zombodze 76% responded that family labour was adequate for all agricultural activities and a similar proportion (70%) in Gege also felt that family labour was adequate. Only 19% in Zombodze and 26% in Gege said family labour was not adequate for all agricultural activities. These two areas are already suffering from some shortage of land so they feel there is "surplus" labour.

The major crop grown in Gege is maize and some beans and pumpkins, while in Maphilenga maize and cotton are equally dominant. In Zombodze, as in Gege, maize is dominant, but beans and potatoes and some pumpkins are also important. However, meaningful output information was available only for maize.

Table 5.8: *Maize Output, 1988/89*

Output (bags)	Gege No.	%	Maphilenga No.	%	Zombodze No.	%	Total No.	%
1 - 5	11	20.4	13	35.1	11	19.0	35	23.5
6 - 10	18	33.3	4	10.8	13	22.4	35	23.5
11 - 15	5	9.2	3	8.1	9	15.5	17	11.4
16 - 20	4	7.0	1	2.7	3	5.2	8	5.4
21 - 25	0	0.0	2	5.4	4	6.9	6	4.0
26 - 30	3	5.6	1	2.7	5	8.6	9	6.0
31 - 35	2	3.7	1	2.7	2	3.4	5	3.3
35 +	3	5.6	3	8.1	5	8.6	11	7.4
No response	8	14.8	9	24.3	6	10.3	23	15.4
Total	54	100	37	99.9	58	99.9	149	99.9

CROP YIELDS

The majority of the respondents harvested between one and ten bags of maize in each area. One bag weighs 90kg. These are small quantities that were used almost exclusively for food. Only 10% in Gege, 21% in Maphilenga and 18% in Zombodze reported that they sell some of the produce. Even then, the produce does not provide sufficient food to last till the next harvest for 45% in Gege, 50% in Maphilenga and 55% in Zombodze. For those who sold agricultural produce, the income each farmer earned was reportedly less than E 1000 except for Maphilenga where 10% obtained up to E 3000 from the sale of agricultural produce including cotton.

Farmers in the three areas generally have to buy maize or mealie meal (maize flour) for food to supplement their produce: 61%, 78% and 71% in Gege, Maphilenga and Zombodze respectively had to buy maize/meal. Only 26%, 14% and 21% respectively did not buy maize/meal, implying a very low level of food self-sufficiency. For the majority of the respondents, the quantities bought varied from one to ten bags of maize or meal equivalent (52% in Gege, 48% in Maphilenga and 55% in Zombodze); and another 3% and 16% in Gege and Maphilenga bought up to 20 bags. While in both Gege and Zombodze no farmer bought more than 20 bags, in Maphilenga another 19% bought up to 30 bags of maize annually to supplement food produced. The buying of maize or meal is done every year once the harvest is exhausted. The cost of the maize/meal is slightly higher than the price they get for selling maize during the harvesting season (table 5.8).

The question on general trends in maize production since 1984 show a high proportion of the respondents (37%, 35% and 31% respectively) saw the general trend in maize production as decreasing, and 24%, 5% and 22% respectively saw the trend as increasing. A significant proportion (20%, 30% and 24% respectively) saw the trend as steady or fluctuating. Only 16% in Zombodze were satisfied by the trend in maize production, and the rest in the three areas did not respond.

LIVESTOCK AND RANGE MANAGEMENT

Most respondents had decided not to answer the question as to whether or not they owned cattle. When the same question was repeated by asking them to state the number of head of cattle possessed by them, the reluctance to respond was similar to that of the previous question except for Maphilenga where only 18% did not respond compared to 35% in the earlier question. In Gege 35% did not respond (previously, 31%) and in Zombodze 34%(previously 45%) did not respond. None of the remainder in the three areas said that they did not have cattle. However, of the majority (48%, 47% and 50% respectively) each owned from one to ten head of cattle; another 11%, 27% and 7% owned up to 20 head; 3% in each area owned up to 30 head each; and only 2%, 5% and 5% respectively of each area owned over 30 head of cattle. Maphilenga clearly stands out as a livestock raising area compared to Zombodze and Gege.

It is common practice in Swaziland, as in many other African countries, to lend out to relatives or to borrow from relatives some head of cattle. In Swaziland this practice is known as "*Kusisa*". In the three study areas 59%, 73% and 38% of the respondents respectively each lent out one to five head of cattle, and only 4% in Gege and 3% in Zombodze lent out up to ten head of cattle. Another 9% in Zombodze however, reported that they loaned out up to 30 head of cattle.

When asked whether they also borrow other peoples cattle, 67% in Gege, 40% in Maphilenga and 43% in Zombodze each borrowed one to three head of cattle and the rest did not respond.

The reasons for practising this "SISA" system are obviously primarily to minimize risk of loss should anything happen. But of the very few respondents who answered the question (76%, 68% and 76% did not respond), the reasons given included convenience (4%, 3% and 3% respectively); no herd boy (7%, 11% and 5% respectively); no immediate need for the cattle (7%, 14% and 12% respectively), and the animals that remain are the adequate number to manage (6%, 5% and 4% respectively). As is usual under this system the "SISA" cattle benefit both the owner and the borrower, the owner reduces pressure on his range resources and also gets alternate calves from the borrower, while the borrower acquires alternate calves and uses the milk. Ownership of cattle on a small scale is spread out in this way.

Too few respondents to warrant meaningful analysis and discussion answered questions about the number of cattle sold and the income generated from the sale. But it is known that cattle are a store of wealth or a bank on hoofs. In times of need for cash for school fees, improved housing, agricultural inputs, buying food, social obligations, etc, cattle are sold. A few farmers sell cattle each year or on a regular basis; 7% in Gege, 14% in Maphilenga and 41% in Zombodze each, sold from one to five head of cattle in 1988/89 season, and 15%, 24% and 1% respectively said they sold none, while the rest did not respond. The average price of cattle was E400 to E700 per animal, hence a possible of income of E2000 to E3500 for those who sold five head of cattle.

Cattle are by and large grazed on communal land (39% in Gege, 32% in Maphilenga and 34% in Zombodze). Respondents who grazed their animal in the mountains were few; 4%, 11% and 3% respectively in three areas. These mountain grazing areas are also communally owned or are public. Grazing on the individual farms was practiced only in Gege (2%) and in Zombodze (3%), while another 3% in Maphilenga and 4% in Zombodze allowed animals to feed on the crop residue and stalks after harvesting. Over 54% in each area did not respond. How are the grazing areas managed and by whom? The majority of respondents, (33% in Gege, 68% in Maphilenga and 26% in Zombodze) said that grazing areas are not maintained by anyone.

Generally therefore, people would try to put in as many animals as they could in such areas as it costs them neither money nor responsibility. Overstocking, resource degradation and soil erosion are evident consequences especially Gege and Zombodze. However, in these two areas with increasing population pressure and land scarcity, 15% and 21% of the respondents respectively practice "community rotational grazing", which has been made possible though some fencing and regulations through the RDAP programme. A small proportion of respondents, 7%, 5% and 3% in Gege, Maphilenga and Zombodze respectively, managed grazing areas that they own as part of their farms.

Ideas which have been suggested at different times in an attempt at proper management of the communal range include possible introduction of a grazing fee to be managed by the community, fencing and more widespread rotational grazing, and decreasing stock numbers. But except for the RDA areas these ideas generally have not been implemented. Communal grazing areas therefore require some urgent attention in the form of controlled grazing coupled with improvement of the grass and shrub varieties, decreasing stock numbers to conform to the livestock carrying capacity, the creation of a contributory communal management fund to pay for some of the labour and other contributions of these directly changed responsible for implementing proper range management measures. Parcelling out the communal grazing land to individual ownership has not been a popular concept, either with the government or with the traditional authorities. While this could lead to better management of both range resources and the livestock economy generally, it could also lead to the cultivation of the (marginal) rangelands under conditions of increasing scarcity of arable land.

Most respondents, 59% in Gege, 68% in Maphilenga and 55% in Zombodze were satisfied that the size of the cropland they were operating was adequate for their needs; but the rest felt that the size of cropland they have could not satisfy their needs. While 4% in Gege, 11% in Maphilenga and 16% in Zombodze asked for more land from the chief, another 8%, 3% and 2% respectively borrowed (leased) other people's fields in an attempt to satisfy their needs for more cropland. However 30% in Gege, 8% in Maphilenga and 26% in Zombodze said there was nothing they could do to remedy the shortage of cropland. 59%, 78% and 57% respectively did not respond to the question of what steps they had taken to satisfy their need for more crop land.

Would the cropland they now possess be sufficient for their children's needs? Yes: 52% in Gege, 51% in Maphilenga and 16% in Zombodze. This land would not be sufficient to meet the needs of their children for 39% in Gege, 38% in Maphilenga and 76% in Zombodze. The responses to the two questions on land adequacy clearly reinforce earlier findings that Zombodze and Gege were experiencing definite population pressure and land scarcity compared to Maphilenga. The children of the respondents will inherit the cropland of the majority of the respondents (i.e. their parents): 78% in Gege, 81% in Maphilenga and 83% in Zombodze.

Usually it is the sons, particularly the first-born sons, who inherit the land. The subdivision of the cropland to the sons has led to land fragmentation and to the reduction of the operation unit to uneconomic sizes of 0.2 to 0.6 ha for some young families particularly in Zombodze. Some of the children obtain land from the chief, and a few through marriages.

Although grazing land is communally owned, the respondents were also asked whether it was adequate for their current needs: 43% in Gege, 57% in Maphilenga and 34% in Zombodze said it was adequate; 19%, 8% and 19% respectively said it was not sufficient for their needs; the rest did not respond. This conforms to the other findings that Gege and Zombodze were suffering from more serious land shortage compared to Maphilenga. In Gege 24% of the grazing land was fenced (46% was unfenced) and in Zombodze 62% of the grazing land was fenced (11% was unfenced), compared to Maphilenga where only 18% of the grazing was fenced. The shortage of land in Gege and Zombodze has made communities in the two areas institute proper management of the communal rangelands, as fencing is usually accompanied by rotational grazing.

Fuel-wood is the major source of household energy in rural Swaziland, and in the study areas, the main sources of fuel-wood were the communal grazing areas (31%, 27% and 48% respectively); the homestead area (15%, 43% and 35% respectively).

Both Gege and Zombodze have forest plantations or forest reserves nearby and these are major sources of free fuel-wood. Settlements in Maphilenga are relatively recent so the natural trees and shrubs, some around the homestead area, are the major sources of fuel-wood.

The majority of the respondents (91%, 95% and 67% respectively) were satisfied with these sources of fuel-wood as being adequate for their current needs. For Zombodze the 33% who thought that these sources of fuel-wood did not meet their demands recommended that more land should be allocated to them (from the private individual tenure farms which seem not to be used) and that more people should plant trees next their homesteads.

CONCLUSIONS

The most common land use management systems in Swaziland are maize cultivation and cattle raising. Maize is both a food crop and a cash crop, and respondents could not state definite quantities of the crop reserved for food or sold for cash. This has made it impracticable to determine the amount of land required for food and for cash crops under the carrying capacity model outlined in chapter two. It has been shown that over 60% of the farmers in each area have to buy maize/meal to carry them over to the next harvest season. Attempts to find out the frequency with which certain types of food are consumed proved futile as most respondents either were not sure or did not respond to a question deemed unnecessary.

Cattle are grazed on the communal rangelands, which were not adequately managed. There is no definite trend where owners sell and derive specified income from their herds. Some of the information collected was not quite reliable. Also the proportion of food provided from livestock (milk and meat) could not be ascertained. Just as in the case of crops, it was not possible to determine meaningful amounts of land required for grazing per family.

Under the circumstance it was felt that the calculation of population carrying capacity of land should not be done. However, since some information is available aggregated at a national level, future researches should attempt the calculation of generalized carrying capacity at the national level using the ecological zones as base units.

Having assessed the soils and land suitability, the chapter found out that farmers generally allocate their crops to land units they had intuitively identified as of the correct suitability. There was no serious "mismatch" of land use: red and black loam soils ("sibovu") were correctly identified as suitable for maize and cotton, and stony soils and steep slopes (>12%) were identified as unfit for cultivation but suitable for grazing. Generally therefore, farmers, through their accumulated experience, practise land evaluation and use planning, and by and large put land units to their suited uses.

Interventions through land use planning by the government would probably endorse current land uses, but most importantly should integrate conservation and resource management measures into the land use plans. This integrated land use planning is necessary not only to prevent further soil loss (Gege and Zombodze) but also to institute measures that will maintain soil fertility under increasing intensive use of land (Zombodze). Only 30% of the respondents in each of the three areas used animal manure and another 30% in Zombodze, 6% in Gege and 3% in Maphilenga used chemical fertilizers. These are rather low levels of inputs, the effects of which are reflected in low productivity, 10 bags of maize per ha as opposed to a potential output 50-60 bags per ha with proper use of fertilizers. Other land management practices include grass strips on contour bunds, some crop rotation, and rotational grazing in the RDAP areas (Zombodze and Gege). These land management measure are not adequate for sustainable production, hence the need to institute integrated land use planning for both arable and grazing areas.

For Swazi nationals, access to arable land is free and equitable under the traditional (Tinkhundla) authority system. Buying and selling of land is forbidden under the law. Once someone is granted arable land someone is free and has the right to use it and pass it on to one's children – a usufruct practice that has evolved to resemble individual freehold systems.

Contrary to findings from other researchers, this study established that the traditional usufruct system did not constitute a direct hindrance to making permanent investment and improvements on the land. Respondents did not express any fear of losing their land, and pointed out lack of capital and lack of water as the major hindrances to permanent improvements on their lands.

However, the communal ownership and use of grazing areas has revealed "the tragedy of the commons" syndrome where everybody has a right to use the communal range and nobody has real responsibility to ensure its proper management. The result has been overstocking far beyond the carrying capacity, resource degradation and soil erosion. The RDAP instituted fencing and rotational grazing measures (practised in Zombodze and Gege) but not all communal grazing areas practise this, and more seriously, it has led to land degradation where decreasing stock numbers or annual off-take rates are not adhered to.

The real problem with Swazi land tenure lies in the national distribution of land: only 52% of the nation's area is Swazi Nation Land – SNL (smallholders including communal grazing range) and sustained 70% of the national population in 1986; compared to 47% of the land in large scale individual tenure land – ITL with less than 10% of the national population. Under conditions of increasing population and land scarcity, the need to acquire ITL land and redistribute it to SNL has been recognised since the early 1970's but land so transferred (bought by the government) has not been redistributed in a clearly known and acceptable manner. This is a national political issue that different governments in Swaziland have tried to solve with rather limited success. Meanwhile most Swazis, including the majority of the respondents in the three sample areas, expect that problems of land scarcity can and will be solved by transferring land from the ITL to SNL system and by establishing viable modalities of how this transfer and redistribution can be effected.

Food production, especially maize, has been decreasing is Swaziland and in the three sample areas the majority of the respondents, 67%, 78% and 71% respectively had to buy maize/meal before the next harvest. The majority in Gege and Zombodze bought small quantities of up to 20 bags annually, but in Maphilenga the majority bought more than 30 bags. Clearly the agricultural output could not sustain the farmers. As for livestock, less than 50% of the respondents in each area owned up to 10 head of cattle each, although in Maphilenga some of the farmers owned more than 30 head each. However, the sale of livestock was not regular and the contribution of cattle income to family subsistence could not be ascertained.

One of the reasons therefore why respondents sought off-farm wage employment, was to supplement agricultural output for subsistence. Almost every family (81%, 78% and 76% respectively) had at least one member of the family in wage employment. In most cases that member was the father (head of the household) or a son, but also mothers and daughters were in wage employment. Most of the wage earnings (79%, 62% and 69% respectively) were used for family subsistence and for agricultural activities. In the economics of household subsistence, earnings from agriculture were becoming of decreasing significance compared to those from wage employment.

The scarcity of land was the second reason why people sought off-farm employment. Over 50% of the respondents in each area had said that they themselves do not hire labour for agricultural activities on their farms because they did not need such hired labour which was generally available and affordable. It would seem logical to accept the hypotheses that the national land tenure system has created land scarcity and landlessness on SNL thus necessitating off-farm wage employment for subsistence. And it was obvious that off-farm employment was not a contributing factor to low agricultural productivity as this labour was in excess of what the small farm units could absorb.

The overall conclusion therefore is that while it is necessary to ensure land capability and suitability and also draw up integrated land use plans for all areas in Swaziland, land scarcity requires urgent attention. Although it is not fashionable to advocate land reform in the age of neo-liberalism, this study reveals that the national land tenure system requires serious transformation, including land reform. One way to ensure equitable food production and distribution, land ownership has to be shifted from Individual Tenure Land (ITL) to Swaziland National Land (NSL) for resettlement after an integrated land use planning has been introduced.

REFERENCES

Allan, W. (1965) *The African Husbandman*, Edinburgh: Oliver and Boyd.

Armstrong, A. 1985. Legal Aspects of Land Tenure in Swaziland. Mbabane.

Blaikie, P. M. and H. C. Brookfields (1986) *Land Degradation and Society* – London Methuen.

Bootsman, C. S. and C. F. Schmidt (1983) A Spatial Analogue Model of Potential Desertificiation in Swaziland. *Desertification Control*, 9:31-39.

Boserup, W. (1965) *The Conditions of Agricultural Growth; The Economics of Agrarian Change Under Population Pressure*. Aldine, Chicago.

Central Statistical Office, (1972) *Agricultural Sample Census (SNL)* 1971 - 1972. Mbabane.

(1978) *Annual Survey of Swazi Nation Land 1971/1972 – 1976 – 1977*. Mbabane, Swaziland Government.

Central Statistical Office, (1980) *Report on the 1976 Swaziland Population Census*. Mbabane, Swaziland Government.

Central Statistical Office, (1986) *Swaziland Census of Agriculture 1983 – 1984*. Mbabane, Swaziland Government.

De Vletter, F. (1983) *The Swazi Rural Homestead. A Case Study of Subsistence, Wage Dependency and Misguided Rural Development*. Cape Town. Carnegie Conference Paper 285.

Fair, T. J. D., and others, (1969) *Development in Swaziland*. Johannesburg: Wits University Press.

FAO, (1986) Food and Agricultural Organization of the U. N. 1986 *Natural Resources and the Human Environment for Food and Agriculture in Africa*. Rome: FAO (Environment and Energy Paper No. 6).

Food and Agriculture Organization of the U. N. (1985) *Project Brief: Swaziland Rural Development Areas Programme*. Rome 112/85 GV – SWA. 9PB.

Food and Agriculture Organization of the U. N. (1978)*Report on the Agroecological Zone Project Vol. 1 Methodology and Results from Africa*. World Soil Resources Report 18. Rome.

Fowler, M. H. (1980) The Approaching Crisis: Population, land and agricultural production in Swaziland, *Report of National Symposium on Population and Development*. Mbabane: Ministry of Agriculture and Cooperatives, pp, 173 – 192.

Hance, W. A. (1970) *Population and Urbanization in Africa*. New York: Colombia University Press.

Hughes, A. J. B. (1972) *Land Tenure, Land Rights and Land Communities on Swazi Nation Land*. Durban: University of Natal Institute of Social Research.

Hunting Technical Services Ltd. (1983) *Review of the Rural Development Areas Programme. Final Report and Annexes: Mbabane: Ministry of Agriculture and Cooperatives*.

Malthus, T. (1978) *The First Essay on Population*: London.

Maro, P. S. (1988) Agricultural land management under population pressure, the Kilimangaro experience, Tanzania, *Mountain Research and Development*, Vol 8, No. 4 pp. 273-282.

Maro, P. S. (1975) *Population Growth and Agricultural Change in Kilimangaro* 1920-1970 University of Dar es Salaam: Bureau of Resource Assessment and Land Use Planning, Research Paper No. 40.

Murdoch, G. (1968) *Soils and Land Capability in Swaziland*, Mbabane: Ministry of Agriculure.

Natural Resources, Land Utilization and Energy Ministry (1986), *Swaziland National Physical Development Plan*. Saad Yahya Associates, Sir Alexander Gibbs and Partners (Africa).

Newman, J. L. and G. F. Marske 1984 *Population Patterns Dynamics Problems*. Prentice-Hall: England Wood Cliffs.

Nkambule, Noah (1988) "The National Land Use Plan Project" Ministry of Agriculture and Cooperatives, Mimeo.

Nsibandze B. M. (1987) History of Soil Conservation in Swaziland, in Peoples Participation in Soil and Water Conservation. *SADCC Soil and Water Conservation and Land Utilization Programme Seminar*. Maputo, 2 –6, 1987 Report No. 10.

Proctor J. D. (1988) "Growth Beyond Limits?" Resource Sufficiency in Rural Swaziland, 1974 to 2000: (Mbukwane-Maseyisini area). M. A. Thesis in Geography. University of California at Berkeley.

Reij, C. (1984) *Back to the Grass Strips*. An analysis of Soil Conservation Policies in Swaziland. University of Swaziland. Mimeo.

Richard P. (1983) "Ecological Change and the Politics of African Land Use". *The African Studies Review*, Vol 26, No. 2. June 1983. (UCLA),

Roder, W. (1977) *Environmental Assessment Swaziland Rural Development Areas Programme. Final Report*. (Contract No. AID 690 – 003 – T) University of Cincinati and University of Zambia.

Russel, M. (1985) *A Report on Landlessness in Rural Swaziland*. Draft presented to FAO

Swaziland Government: (1986) *Report on the 1966 Swaziland Population Census*. Mbabane. Swaziland government.

Swaziland Government:: (1984) *Fourth National Development Plan* 1983/84 – 1987/88. Mbabane. Swaziland Government.

Swaziland Government: (1984) *Project Report of the Rural Development Areas Programme Phase III. Annex 1: Sub-Areas Description*. Mbabane. Ministry of Natural Resources, Land Utilization and Energy.

USAID, (1980) *Environmental Profile of Swaziland*. Library of Congress: Science and Technology Division, Aid Control No. SA/TOA 1 –77.

6 Price and Market Liberalisation Effects on Household Food Security: A Comparative Gender Sensitive Study of Rural and Urban Uganda

Henry Manyire

In mid-1981, Uganda introduced stabilisation and Structural Adjustment Policies (SAPs), which were quickly engulfed by political unrest and instability. In 1984, SAPs were abandoned due to the intensification of the civil war. In May 1987, the National Resistance Movement (NRM) government took the reign of power and began an economic recovery programme, aimed at restoring economic stability, establishing more realistic relative prices and rehabilitating Uganda's productive and social infrastructure. Several policy measures have since been implemented, including price and market liberalisation.

According to SAPs, price liberalisation refers to the removal of government subsidies on consumer goods and capital equipment such as farm inputs, dismantling of price and distribution controls, and bringing domestic prices into line with world market prices (less transportation and marketing costs). Market liberalisation, on the other hand, refers to removal of state own enterprises (parastatals) monopoly over procurement and marketing and the lifting of export/import restrictions. Both policy instruments stress increased reliance on market forces and the private sector, with the concomitant withdrawal of the state from resource allocation. The overall goal SAPs is to correct distortions in the structures of the economy such that ultimately, demand and supply forces determine the prices of goods and services leading to efficient resource allocation.

However, in the short run, the economic policy reforms were a source of controversy, particularly the questioning of their economic and social benefits as well as their potentially negative consequences on household food security. There appears to be a contradiction between macro forecasts and micro impacts. Notwithstanding the economic policy reform's controversies and critique, this chapter contents that SAPs micro-economic impacts depend on how a given community restructures its resources and livelihood possibilities. In the Ugandan context, as elsewhere in the developing countries, there are structural differences between rural and urban households' accessibility to food, and there are also differences amongst the rural communities. Whereas urban households have a high recourse to market purchases to meet their food needs, rural subsistence farmers provide some of their own food requirements. Rural households dependent on wage employment may also depend on market purchases, although some may be paid

M.A.M. Salih (ed.), Local Environmental Change and Society in Africa, 2nd Edition, 129–158.

predominantly in food. Alternately, there are some rural commercial farmers who produce exclusively for the market and they too, may rely on market purchases. And, although households provisioning of their own food in the good years may shield them against bad years, they may be driven into a hostile food market by unpredictable drought shocks.

Besides the mode of accessibility to household food, there are also differences in consumption patterns. Whereas rural households depend on non-tradable (tubers, roots, vegetables, and plantains), most urban households consume tradable (maize-flour, beans, groundnuts, rice) which are more affected by price and market liberalisation for they are exportable. Given such structural differences in modes of household accessibility to food, production, consumption, and expenditure patterns, the effects of price and market liberalisation on the availability and affordability of household food are likely to vary.

In addition to structural differences in accessibility to food, the effects of price and market liberalisation may be gender differentiated. For although the market does not always operate against the interests of women, in the context of adaptation to rapidly evolving external economic conditions, the dual productive roles of women within national economies take on great importance. Available time after household provisioning and maintenance and differential access to economic opportunities and resources constitute the major constraints to the expansion of female accessibility to household food. Furthermore, as substitute service providers and mangers of the domestic economy, the burden of increased costs of food due to price and market liberalisation falls largely on women.

However, it ought to be noted that women are not a homogeneous group. For their accessibility to food more especially in rural areas depends not only on where they live, but also on their place within the rural economy; ladles or land owning, tenant farmers or share croppers, members of a co-operative or communal farm, size of land holding, ownership of won plot etc. similarly, in the urban areas, women may be either entrepreneurs, self-employed, employees, casual workers or unemployed which influences income, and subsequent effective demand for food differently. Socio-demographic variations like martial status, level of education attained, family connection, number of children, and female headship of household influence resource in and outflows, which may further determine accessibility to food.

Therefore, the main objective of this research is to investigate the effects of price and market liberalisation on grassroots households' food accessibility amongst urban and rural men and women in Uganda. It focuses on investigating the ability of urban and rural households to meet target consumption levels of food under the liberalisation era, either through self-provisioning, purchasing or both. Special attention was paid to the food security situation of men and women. By assessing the relationship between Market and Price Liberalisation on one hand, and accessibility to household food, intra-households distribution of income and food and other household socio-economic characteristics on the other hand, we would expose the mechanisms through which the urban/rural household's' food security is affected. Furthermore we can as a result determine the consumption patterns adopted by the households in their struggle to safeguard food security the efficacy of households' responses to the transformations engendered by economic liberalisation.

CONCEPTS AND RELEVANCE

The influence of price and market liberalisation on household food availability/affordability among the various socio economic groups in rural and urban Uganda would depend on the nature of the markets involved, the institutional setting of the intermediate economy, and the characteristics of the households in each group. There are locational (rural/urban) intra locational variations in the shares represented by exportable and other non-food tradable products (coffee, cotton, vanilla), tradable staple foods (maize, beans, millet, groundnuts), and non-tradable foodstuffs (roots, tubers, plantains, vegetables) in the production and consumption patterns.

A number of studies have been conducted on the impact of adjustment on the poor in general, and on the effects of price and market liberalisation on household food security in particular. Pistrup-Andersen, (1988), Cornia *et al,* (1987) and World Bank 91990), argued that since the rural poor were often net consumers of staple grains, especially traded goods whose prices are assumed to increase commensurate with price and market liberalisation, adjustment programmes were asserted to have deleterious consequences for the rural poor.

Tokman (1989) observed that structural adjustment policies enacted in response to the recent economic crisis have especially affected poor people, many of them in the urban areas. Aboyade (1988) also noted that adjustment programmes concentrated on improving the balance of payments and managing debts rather than protecting the vulnerable and impoverished, among whom women and their children predominate. Massiah (1986) further added that structural adjustment programmes undermined attempts to ensure and maintain and adequate livelihood for women and those in their care, especially through increasing urgent need for cash to buy basic necessities, but she was quick to caution that sociological issues were also involved.

Massiah, (1989), Elson, (1987), Strassman (1987), Le-franc, (1990), Jiggins (1989), O'brien, (1985), Cornia *et al* (1987), Mutemba, (1989) and Zuckerman, (1989) further elaborated that the extent and type of women's involvement in earning was not only related to the national level of socio-economic development, but also to the prevailing gender ideology and women's personal circumstances. They asserted that socio-economic conditions continued to restrict women's participation in the economy, limited their mobility, and ignored the deleterious effect on women of the marco-development strategies of the 1980's. the authors added that women were faced with growing unemployment, declining real income, and food insecurity, among other problems, all of which were not gender neutral problems. They were problems affecting women, rather than men, or to a greater extent than men, for two separate but interconnected reasons. One related to women's separate but interconnected reasons. One related to women's reproductive and caretaker roles, and how they are perceived, the other related to the mechanisms provided by prevailing development strategies for women to carry out those roles. In support of this, Leslie et al, (1988), Boyle, (188), Berger, (1988), and Carena and Sebstad, (1989) added that heightened unemployment and downward pressure in wages for men under structural adjustment programmes had meant that women'' cash income, although earned in menial wage work or from micro-enterprise activities, had taken on great importance in meeting the basic needs of the household.

On the other hand, for the increasing number of female-headed households inmost parts of the world, the general fall in wages and formal sector employment opportunities for women had resulted in a significant decline in their economic status, plunging many into absolute poverty. The authors were of the opinion that the interaction of class and gender factors in trapping the most disadvantaged women in poverty was only partially mitigated by vertical solidarity.

On food security, Boyle, (1988) intimated that the reduction or elimination of food subsidies had potentially dire consequences for the poorest of the poor, who may infect, have been quite numerous in relative and absolute terms even in middle-income countries. World Bank (1986) noted that up to 36% of the population of Morocco would fall beneath the level of nutritional adequacy with the desubsidization of the price of bread wheat. Boyle (1988) further added that consumption strategies designed to safeguard food intake and other basic necessities as household income falls or as food prices rise, were devised and managed by women as part of their household provisioning responsibility. Common adaptive behaviours were reported to include: changes in purchasing practices in order to reduce necessities; changes in the types of goods and services consumed, particularly ture of food times; modification of dietary patterns: and deliberate altering of intra-household food distribution.

Cornia *et al* (1987) also noted that the most critical changes in household consumption patterns involved changed dietary practices. They noted that poor households that already spent form 60% to 80% of their total income on food "wee forced", first to increase the proportion of food expenditure in total expenditure; secondly to concentrate their food expenditure almost exclusively on calories; and thirdly, to substitute cheap for expensive sources of calories.

A contrary point of view however, is that adjustment programmes stand to help the rural poor directly or indirectly by raising producer incentives and rural incomes. Relative price movements commensurate with adjustment were seen to favour products produced by small holders whose real incomes consequently increased. (Whyte, 1984, Belshaw, 1988, Bates, 1981 Buker, 1985, Adesini, 1981, Choi, 1987). Similarly, Carena and Sebstad (1989) and ILO/JASPA (1981) suggested that measures such as those that decreased competition with foreign goods or increased the availability of foreign exchange, and liberalised prices and product markets served to expand employment and income opportunities for women although implications for household food were not mentioned.

Addison and Demery, 9185), Sahn, (1990), Sahn and Saris, (1991), and World Bank, (1990) gave a more balanced and acrostic view points; that the impact of adjustment on the poor was not knowable apriori. Poverty in the Bank's view, would be determined by the propensities of the poor to produce and consume tradable versus non-traded as well as by the actual movements in the relative prices. Commonwealth Secretariat (1989) supported these view points with illustrations from Zambia where changes in the prices of tobacco and groundnuts compared with maize led to a widespread decline in the production of these cash crops as farmers shifted to maize production.

However, there was neither evidence of female accessibility to income from maize (a man's crop), nor evidence to suggest that the additional income was spent on household food requirements. Blumberg, (1988), similarly noted that ironically, the growing shift of family labour into cash crop production and, or, cash wage work sometimes failed to result in a proportional rise in household welfare. Studies show that adult male household members often held back a higher proportion of their income of personal consumption than did their wives. It is also possible that men may appropriate family income from whatever source, particularly the head of household for personal use. Certainly, much more needs to be know about income disposal and sharing of various wage earners within the household, in relation to food purchasing and distribution.

Conversely, Makgetla, (186), and Manyire, (1992), indicated that there were no correlation's between increased price incentives and agricultural output in Zambia and Uganda respectively, particularly among the smallholder farmers. This was attributed to the limited supply elasticity of the peasant farmers, for high prices did not necessarily increase profits significantly especially if infrastructure was poor or when input prices also rose in the case of Zambia, and farmers did not feel the price incentives via effective demand for their produce, in the case of Uganda. However, in Uganda, Manyire (1992) noted that production of non-tradable foodstuffs was stable, although the correlation between stability of non-tradable production, and food security was not investigated which called for further research.

The above reviewed literature indicated that most of the previous findings, especially Zambian studies had been contradictory; whereas some studies indicated a positive response to adjustment, others showed a negative response while some were neutral. Further, most of the studies seemed to have concentrated on the relative quantity supply responses, with little attention paid to specific income changes and virtually no focus on availability/affordability of household food.

Most studies had documented the impact of policy reforms from only macro-economic perspectives the manner in which local initiatives for self reliant development were affected is less know. Little empirical work has been done at the household level to determine either the precise ways in, which the lives of various groups have been affected; or the ways in which these groups have attempted to adapt to changing conditions under adjustment. Where household assessments have been made, the information was analysed in terms of a macro-policy context, presented mainly as descriptive and anecdotal studies.

Furthermore, most studies tended to treat women and men as having similar level of access to productive resources, yet in reality women are constrained and their access to input is consequently disadvantaged compared to men's. As a result, the policy relevant questions relating to how the returns of different gender's employment (self or otherwise) in rural and urban areas could be increased in order to enable poor households to meet more effectively their food needs were not raised. Studies that have attempted a gender specific economic analysis of the impact of adjustment were too general and highly conjectural which suggested a need for more in-depth empirical studies. However, the studies were of the view that adjustment affected the rural and urban poor's production and consumption patterns, although the actual modalities of such influence, and, or, responses were not considered.

THE RESPONDENTS' SOCIO-DEMOGRAPHIC CHARACTERISTICS

The fieldwork results on which this chapter is reveal that the urban areas attracted the younger generation much more than the rural Agricultural Sector. Whereas 63.9% of the respondents in the urban areas were aged 30 years and below, in the rural agricultural sector, this age category comprised 37.4%. This differential could be attributed to the declining agricultural terms of trade, and falling living standards endemic in the rural areas during the last two decades. Consequently, majority youth shun the rural areas in general, and agriculture in particular, opting to sell their labour in the urban areas, especially in the urban informal sector. In this sector, activities therein are regarded by the youths as more profitable, promising and easier than agriculture whose products' falling prices were discouraging the young farmers (Manyire, 1995).

With regard to educational level attained, only 8.0% in rural areas had completed 7 years of schooling compared to 22.0% in the urban areas. Seventeen percent in rural Uganda had more than 7 years of formal education compared to 40% in the urban areas. Rural areas had a high population of respondents who had never gone to school, 42.4%, compared to only 12.0% in the urban areas. The large disparities in educational attainment between the urban and rural areas is due to the decline in the real value of Government financing of the social sectors during the long years of economic deterioration. Hence, the burden of financing primary schooling in Uganda had shifted heavily to parents, with the result of "the lower income groups' especially in the rural areas either not enrolling, or withdrawing children form school when the finical burden becomes too great" (World Bank 1993a).

In terms of marital status, majority respondents were married. Rural Uganda had the least respondents in the single marital status category – 7.1% compared to 18% in urban Uganda. This is indicative of the tendency of marrying at a much younger age in the rural areas compared to the urban areas, in part due to the high drop-out rates (Republic of Uganda, 1993) and partly due to the lower cost of living in the rural areas compared to the urban areas.

It is therefore not surprising that rural Uganda had the least reporting of respondents who had no children, 9.1%, compared to Urban Uganda where 21.0% reported not having children. Majority respondents had 1-5 children, although in the case of rural Uganda 43.4% had between 6-10 children. This is testimony of the higher fertility rates in Uganda, which average 7.3 births per woman (Republic of Uganda, 1989).

With regard to household size, Uganda urban exhibited the highest incidence of small household sizes; 58.0% had household sizes of 1-5 people. Rural Uganda had 50.5% respondents reporting that they had household sizes of 6-10 members. Further 26.3% in rural Uganda reported household sizes of more than 10 members. Most of the households with large household sizes had dependants, other than own children, indicative of the large prevalence of the practice in Uganda of urban residents sending children to the rural areas to be looked after by relatives therein, when the cost of living in the urban areas becomes unaffordable.

CHANGE IN HOUSEHOLD FOOD CONSUMPTION PATTERNS

Change in household food consumption patterns were measured in terms of changes in number of meals taken per day over per the past six years, and changes in the type and quantity of food taken per day over the same time period. The past six years time span incorporates the period prior to, and during structural adjustment, under which price and market liberalisation policies have been implemented.

The norm in most Ugandan households is to take a minimum of three meals per day. These are breakfast, lunch and supper. Evening tea is largely an urban phenomenon though a few rural households, which have spent some time in urban locations have adopted the practice. The number of meals taken by the respondents per/day are indicated in table 6.1.

Table 6.1 shows that urban households conformed to the norm of having a minimum of three meals per day more than rural households. Fifty per cent of the urban resident reported having a minimum of three meals a day compared to 37.3% of their rural counterparts. Alternatively majority rural respondents – 57.5% reported having 2 meals a day compared to 42.% of their rural counterparts. The two meals were mainly lunch and supper.

Changes in the number of meals taken by households per day over the past six years are shown in table 6.2.

Table 6.1: *Meals taken by households per day by location by gender location*

	Urban		Rural	
Number of Meals	Males	Females	Males	Females
	%	%	%	%
3+	32.0	18.0	22.2	15.1
2	29.0	13.0	33.3	24.2
1	6.0	2.0	1.0	4.0
Table %	67.0	33.0	56.5	43.4
Total (n)	67	33	56	43

Table 6.2: *Changes in number of meals taken by households per day over the past six years by gender by location*

Location	Urban		Rural	
Sex	Males	Females	Males	Females
Changes	%	%	%	%
Increased	12.0	3.0	11.1	4.0
Decreased	18.0	11.0	12.1	15.1
Same	37.0	19.0	33.3	24.2
Total %	67.0	33.0	56.5	43.3
Total (n)	67	33	56	43

Interestingly, 56.0% of the urban respondents and 57.5% of the rural respondents reported that there had been no changes in the number of meals taken prior to, and during the liberalisation of prices and markets, implying that liberalisation had not affected over half of the respondents' the number of meals taken per day. However, 29.0% of the urban respondents, and 27.2% of their rural counterparts reported that the number of meals taken in their households had declined during the liberalisation era. Fifteen per cent in either study location reported that the number of meals taken by their households in the preceding six years had increased. However, it were the males who overwhelmingly reported increases in number of meals taken compared to females, i.e. about three-quarters of respondents in each study location who reported increased in number of meals taken in households were male. Only one quarter were females.

With regard to types of foods consumed by households, Table 6.3 showed that majority respondents in both urban and rural areas consumed non-tradables. These included roots, tubers, plantains, and vegetables. However, urban residents consumed more tradable foodstuffs compared to their rural counterparts i.e., 73% and 36.2%, respectively.

As for changes in the types of foods taken by households over the past six years, Table 6.4 showed that over one half of the respondents in both the rural and urban areas had not changed from the types of food they had been taking prior to adjustment. Implicitly, consumption of non-tradables is the tradition in Uganda.

Table 6.3: *Type of foods taken by households by location by gender location*

	Urban		Rural	
	Males	Females	Males	Females
Types of Foods	%	%	%	%
Non-tradable	62.0	44.0	58.5	53.5
Tradable	46.0	27.0	17.1	19.1
Total (n)	67	33	56	43

Total % is more than 100 because of multiple responses.

However, 27.0% in the urban areas, and 26.2% in the rural areas reported decreases in types of foods consumed, while 17.0% in the urban areas, and 13.0% in the rural areas reported increases in the types of food consumed. Once again more males than females in both the rural and urban areas reported increases in the types of food taken since the liberalization of prices and markets.

The changes in the types of foods taken refer mostly to the changes in the consumption of tradables e.g. sugar, rice, wheat products, and the high cost non-tradable food, especially meat, fish and milk, the main protein sources. Thus, it appears that for the majority households' low cost calorie foods e.g. cassava, potatoes, and vegetables were major substitutes in the wake of price and market liberalization.

Table 6.4: *Changes in types of food taken by households over the past six years by location by gender*

	Urban		Rural	
	Males	Females	Males	Females
Changes	%	%	%	%
Increased	3.0	4.0	9.0	4.0
Decreased	18.0	9.0	15.1	11.1
Same	36.0	20.0	32.3	28.2
Total %	67.0	33.0	56.4	45.3
Total (n)	67	33	56	43

With regard to availability of enough household food at all times, field data revealed that 20.0% in the urban areas and 15.0% in the rural areas answered in the affirmative, implying that majority respondents both in the rural and urban are as faced some form of food insecurity.

FACTORS ACCOUNTING FOR CONSUMPTION PATTERN CHANGES

Food security is a function of the households' socio-economy, which socio-economy is not only largely influenced by income and profitability accruing from self-employment/own production, but is also strongly correlated with welfare. The households' socio-economy includes the provision of, and access to food, other domestic consumables, health and education.

It is interesting to note that this study's findings refuted the widely held beliefs that agricultural households in rural Uganda produce most of their food requirements. This study revealed that majority rural agricultural producers resorted to market purchases of food at some time. Only 13.1% of the rural producers provided on their own all their food requirements.

As expected, only 7.5% in the urban areas reported that they produced all their food requirements without recourse to market exchanges. This particular category had one leg in the rural and, or pre-urban agriculture, and another in the urban employment. Given the high recourse to market exchanges to purchase the food required for household consumption, it was imperative to assess the changes in expenditure on food for household consumption experienced during the preceding six years. Table 6.6 shows that 70.0% in the urban areas and 73.7% in the rural areas were spending more on food for household consumption since the liberalization of prices and markets' policies were implemented. Only 12.0% in the urban areas, and 6.1% in the rural areas reported that they were spending less than during the price liberalization era.

Table 6.5: *Changes in expenditure on food for household consumption by Location*

Location	Urban	Rural
Change	%	%
Increased	70.0	73.7
Decreased	12.0	6.1
Same	11.0	7.1
Not applicable	7.0	13.1
Total %	100.0	100
Total (n)	100	99

Reasons for the changes in expenditure in food for household consumption are indicated in Table 6.6.

Increased prices of food, was the most cited factor associated with food insecurity in urban areas, reported by 66.0%, followed by increased number of dependants, cited by 5.0% of the respondents. Nine percent of the urban respondents had reduced expenditure on food due to a decline in their income, hence were consuming less food than was required to lead an active healthy life. Only 3.0% attributed a decrease in expenditure on food to decreased number of dependants, implying that although there was a decrease in expenditure, this did not compromise their food consumption levels for the decrease was not due to increased prices or reduced income.

Table 6.6: *Reasons for changes in expenditure on food for household consumption by location*

Location	Urban	Rural
Increased prices	66.0	23.0
Increased number of dependents	5.0	8.0
Drought	-	42.7
Decreased number of dependents	3	6.1
Same	9.0	20.2
Not applicable	18.0.0	30.3
Total %	100.0	100
Total (n)	100	99

Source: Field Data

Conversely, drought was the single largest factor responsible for decreases in food consumption in the rural areas, reported by 42.7% of the respondents. This was followed by increased prices of domestic requirements reported by 23.0%. Rural respondents who attributed a decrease in expenditure on food due to a decline in income comprised 20.2% implying that they were consuming less than was required to lead an active healthy life.

However, drought most affected the commercial farmers, 13.1% of whom significantly attributed decreased food consumption to the phenomena. Subsistence farmers least attributed the decreased food consumption - 3.0%, of the commercial-cum-subsistence farmers were of the view that drought was responsible for decreased food consumption constituted 9.1% of the respondents.

The above statistics point to the possibility that rain failure alone does not cause food insecurity. Vulnerability to drought is determined by rural poverty, and trends that influence overall economic productivity, levels of savings, and the diversity of income sources - factors that determine how well a population can cope with drought and other shocks (World Bank 1993c). Thus the institutional setting amongst the commercial farmers was less conducive to cushioning shocks like drought, through resorting to non-agricultural sources of income and livelihoods, for example selling of labour and engaging in off farm activities. Similarly, cropping patterns and choices of mix could also mitigate or exacerbate the effects of drought. For instance, the subsistence sector by virtue of its institutional setting could cushion the effects of drought much better than the commercial sector. While the commercial producers relied mostly on cotton, maize and bean production, the subsistence farmers had a diversified cropping system including yams, cassava, and potatoes, which are drought resistant, and bananas which are both perennial and more drought resistant than the annuals produced by the commercial farmers. Hence, even with drought, the subsistence farmers' food consumption had not decreased significantly.

In the Urban areas, increased food prices largely accounted for decreases in food consumption: 16.7% of the urban workers attributed decreases in food consumption to increases in prices of food. Increases in prices of food in the urban areas were mainly exacerbated by drought in the food producing areas, inflation, and increased transport costs which food traders incurred in transporting food from the rural to the urban areas. That is why rural Uganda least showed that food prices had increased, i.e. only 23% of the respondents.

With regard to changes in expenditure on other domestic use commodities, in both locations the expenditures had increased as indicated in Table 6.7.

Table 6.7: *Changes in expenditure on other domestic use commodities*

Change	Urbam	Rural
Increased	76.0	80.8
Same	11.0	10.1
Decreasd	13.0	9.1
Total %	100.0	100
Total (n)	100	99

As in the case of food, increased prices of these commodities were largely responsible for the increased expenditures. The increase in prices of domestic consumables is largely attributed to liberalization of the foreign exchange markets, devaluation and inflation.

Table 6.8: *Reasons for the chances by location*

	Urban	Rural
Changes		
Increased prices	69	76.9
Increased family size	3.0	3.0
Increased children demands	4.0	1.0
Parents/relatives assets	3.5	2.0
Less production	-	3.0
Not applicable	13.0	9.1
Total %	100.0	100.0
Total (n)	100	99

The education related expenditures were similarly reported to have increased in both locations as shown in Table 6.9.

Table 6.9: *Change in education related expenditures by location*

	Urban	Rural
Changes	%	%
Increased	58.0	59.6
Same	6.0	4.0
Decreased	2.0	3.0
Not applicable	34.0	33.0
Total %	100	100
Total (n)	100	99

However, the proportion of the respondents who reported increased expenditures on education was lower than of those who reported increased expenditures on food and other household/domestic use commodities. This is because there were more households that had no school going members, yet most households purchased food, while all households purchased other commodities for domestic use. Public expenditure cuts under structural adjustment, which involves cuts in allocation of public funds to education were largely responsible for increased household expenditures in education. With cuts in public expenditure, the proportion of public expenditure going to education in 1987 was only 1.6% in Uganda in 1989/90, (World Bank 1993a). Hence, private expenditures on both the education and health sectors far outweigh government spending, which is a notable feature of poverty in the country (Udsholt, 1994). Approximately two thirds of total health expenditures and as much as three quarters of spending for primary education are privately financed i.e. by households! The increases in the education related expenditures imply a trade-off of resources that would have otherwise have been spent on households' welfare, including food consumption.

In terms of health related expenditures, Table 6.10 shows that majority respondents, both rural and urban based had experienced increased expenditures in the wake of adjustment.

Table 6.10: *Changes in health related expenditures by location*

	Urban	Rural
Changes	%	%
Increased	63.0	68.7
Same	8.0	7.1
Decreased	4.0	16.2
Not applicable	25.0	8.0
Total %	100	100
Total (n)	100	99

Most households had suffered an illness in the preceding 90 days, the single largest ailment being malaria. Public expenditures on Health were 0.6% of GDP in Uganda in 1990/91 (World Bank, 1993a). Hence, the Uganda government had heavily reduced its expenditure on health thus passing on the burden to the population. Majority Ugandans used private clinics/dispensaries where services are paid for. It is also important to note that more urban than rural households used government hospitals in Uganda. This is because of two reasons: first, most government health services in Uganda are located in urban areas, and second, urban health units have at least the basic facilities compared to government aided rural health centres which are usually abandoned and lack basic facilities such as simple drugs and equipment e.g. syringes.

Besides own production, food security is also influenced by the household's income. It was therefore necessary to assess changes in income over the previous six years, and the extent to which returns from own account production met both the rural agricultural producers' and urban workers' household food and production needs. In rural Uganda, 47.0% reported increases in income over the preceding 6 years compared to 37.9% in the urban areas.

That over one half of the respondents in Uganda's household agriculture had not experienced increases in income following liberalization lends credence to the near absence of price controls in Uganda especially on the staple foodstuffs, prior to adjustment. The private sector has always been functioning with considerable efficiency to market food crops in Uganda (Republic of Uganda, 1990a).

Hence, even with adjustment, prices did not change considerably, except for the traded commodities especially coffee, whose marketing had been formerly under government control. Table 6.11 shows that subsistence farmers reported highest the increases in farm income.

Similarly, the subsistence farmers least reported decreases in farm income - 4.0% compared to 16.2% of the commercial farmers, and 12.1% of the subsistence cum commercial farmers. Reasons for the variations in changes in farm income amongst the agricultural producers of the different orientations are indicated in Table 6.12.

Table 6.11: *Changes in farm Income over the past six years by agricultural orientation*

	Commercial	Subsistence	Commercial and Subsistence-
Changes	%	%	%
Increased	9.1	30.8	7.1
Same	28	15.2	3.0
Decreased	16.2	4.0	12.1
Total %	28	49	22

p = 0.00

Table 12: *Reasons for change in farm Income over the past six years by agricultural orientation.*

	Commercial	Subsistence	Commercial and Subsistence-
Reasons	%	%	%
Increased prices	5.1	24.2	5.0
Cultivate more	-	3.0	2.0
Increased prices of domestic requirements	3.0	2.0	
Poor markets	1.0	1.0	-
Drought	13.1	3.0	9.1
Not applicable	6.1	18.2	5.1
Total %	28.3	49.5	22.2
Total (n)	28	49	22

P = .00

The increased income accruing to the subsistence farmers is attributable to two reasons: first, removal of government subsidies on agricultural inputs used by the commercial farmers had sky rocketed their (commercial farmers) costs of production, yet output prices had not correspondingly increased (Manyire 1992). Alternatively, the subsistence farmers relied mostly on traditional tools (hoe, and *panga*), rarely hired labour and tractors, and did not use hybrid seeds, sprays and pesticides, all of whose prices had increased following adjustment. Hence their costs of production were stable, thus returns higher, given the slight output price changes. Secondly, since 1986, there has been an increase in urban demand for food.

With high levels of donor support, there has been an increase in urban incomes - and demand - resulting in part from the refurbishment of industries, services, and government (World Bank, 1993b). Hence, subsistence staple food producers have experienced modest income increases.

With regard to the gender distribution of changes in farm income, there were no significant differences between males and females as indicated in Table 6.13.

Table 6.13: *Chances in farm income over the past six years by sex*

Changes	Male (%)	Female (%)
Increased	29.3	17.2
Same	12.1	9.1
Decreased	15.2	17.1
Total %	56.6	43.4
Total (n)	55	44

P= .44

However, slightly more males than females reported increases in farm income over the past six years. Similarly, slightly more females than males reported that farm income had decreased over the same period. Gender disaggregated reasons for changes in farm income are indicated in Table 6.14.

Table 6.14: *Reasons for change in farm income over the past six years by Sex*

Reasons	Male (%)	Female (%)
Increased Prices	21.2	13.1
Cultivated More	4.0	1.0
Increased Prices of Domestic requirements	4.0	-
Poor Markets	1.0	1.0
Drought	10.1	151.0.2
Not applicable	16.2	13.2
Total %		

Field Data

In the Urban areas, there were no significant gender and employment status category differences with regard to changes in income over the previous six years as indicated in Table 6.15. However, the percentage of males who reported increases in income was more than twice that of the females.

Table 6.15: *Changes in urban incomes over the past six years by sex*

Changes	Male (%)	Female (%)
Increased	27.1	10.8
Same	7.7	8.3
Decreased	38.3	13.3
Total %	67.5	32.5
Total (n)	67	33

P=0.07

Probably, this was because the highest reporting of increases in urban income, 16.7%, was in the overwhelmingly male dominated manufacturing category of the urban informal sector.

Measurement of the extent to which returns from self employment met both the producers'/operators' household food and production needs revealed that only 11% in rural Uganda had their returns meeting both their domestic and production needs. And in the urban areas 35.8% of the urban workers reported that returns from their self-employment met both their domestic and production needs. It is interesting to note that although the percentage of urban workers whose returns from self-employment met both their household food and production needs was more than thrice that of the rural agricultural producers. However, returns similarly met their household food and production needs, more rural producers had reported having experienced increases in income during the preceding six years than the urban workers.

Implicitly, increased incomes per se do not guarantee the meeting of households, food needs, unless the increases are real. In the context of rural household agriculture, the increases in incomes were not commensurate with the increases in prices of goods and services consumed by these rural households hence there were no improvements in welfare. Alternatively, the increases in urban incomes were high enough to cover the increases in the prices of other goods and services consumed by the urban households, thereby leaving a significant amount for both welfare (food) and re-investment (production).

The sex disaggregation of the extent to which agricultural production returns met both the producers' food and production needs is indicated in Table 6.16.

The gender distribution of agricultural producers whose returns met both their food and production needs was not even: 10.1% of the males reported that their returns met their domestic and production needs compared to only 1.0% of the females. This points to the gender asymmetries in the rural agricultural household sectors. Women's lower socio-economic entitlements in the rural socio-economy hinder the realisation of higher production returns, which would

Table 6.16: *Extent to which agricultural production returns met both producers' food and production needs by sex*

	Males (%)	Females (%)
Met them	10.1	1.0
Did not meet them	46.5	42.4
Total %	56.6	43.4
Total (n)	56	43

P = .05

improve both welfare and investment in production: fewer women own land, and their access to credit is much lesser than that of males. So is their command of labour required for production, given the competing demands for labour in the execution of the domestic obligations and responsibilities. Further, females produce predominantly for household consumption while male produce mostly for the market, hence they could have had little encounters with the market. Besides, females have limited control over income from agricultural produce, whose marketing is largely a male preserve. UNICEF (1989) has indicated that while 68% of the labour required for women provided cultivation of food crops in Uganda, while only 30% of them could sell the crop. It is also possible that majority of those who could sell the crop were female heads of household, who coincidentally constitute 30% of all households in Uganda (World Bank, 1993d). That males whose returns from agricultural production were ten times more than females, yet 73% of all the respondents in the rural areas were married, implies that males may not necessarily be spending the increased incomes from agricultural production on household welfare, including food.

There were no significant differences in the extent to which returns met both food and production needs amongst producers of different agricultural orientations as indicated in Table 6.17.

Table 6.17: *Extent to which agricultural production returns met both producers' food and production needs by agricultural orientation*

	Commercial	Subsistence	Commercial and subsistence
Extent	%	%	%
Met them	3.0	4.0	4.0
Did not meet them	25.3	45.5	18.2
Total %	28.3	49.5	22.2
Total (n)	28	49	22

P = .38

In the Urban areas there were no significant gender and employment status differences in the extent to which returns from self-employment met both the workers' food and production needs as shown in Table 6.20 and 6.21. However, a higher percentage of males reported that their returns met both the domestic and production needs, compared to females, although the percentage of urban females whose returns from self-employment met their food and production needs was much higher than that of rural females, i.e. 15.0% compared to 1% respectively.

CAREER PATTERN CHANGES AT THE HOUSEHOLD LEVEL

75.7% of the rural producers produced non-tradable crops (plantains, tubers, roots) market for household consumption, 35.4% produced traded commodities (coffee, cotton) for the market, 51. 5% produced tradable crops (maize, beans, Soya beans, cereals) for both the market and household consumption, while 11.1% did not produce for the market at all. Only 11.1% reported that the crops grown presently for cash sale were not grown in the preceding six years, indicating that there were not many marked shifts in cropping patterns in response to enhanced price and market incentives as subsumed in the factor and product market liberalization theory. However, for the few respondents who had shifted to different crops, improved marketing opportunities and prices for the crops were the major reasons that prompted the shifts. Surprisingly, fewer producers marketed their produce, implying that wholly subsistence agriculture (producing for own consumption) was a key characteristic of Ugandan agriculture.

Moreso, only 10.1% reported having increased area size cultivated in the preceding six years while only 12.2% had decreased the area size cultivated. Majority - 67.6% had maintained the same area sizes.

Field results indicate that better markets and prices, coupled with availability of inputs and labour were less experienced in the rural household agricultural sector, yet these are supposed to be ushered in by the liberalization policies.

54. 1%. Of the respondents reported a decline in employment opportunities. Only 23. 3% indicated that their employment activities had increased. The rest of the respondents maintained that there had been neither increases nor decreases in the volume of employment activities during the past six year. Simultaneously, there has been an increase in the number of workers in the Urban Informal Sector (U.I.S) reported by 59.2%. 31. 7% reported a decrease in the number of workers in the state sector.

Reasons for the changes in number of workers in the sector include the fact that more people were joining the urban informal sector either to wholly work there, or to supplement the eroded real wages in the formal sector where wages could barely sustain a family. The attractiveness of the U.I.S. is further exhibited in the low capital and skills required for entry. Increased profits as an added attraction to the urban industrial sector was reported by only 21.0%. Alternatively, a decline in demand for the U.I.S. goods and services, coupled with a decline in the working capital of the workers were the major reasons cited for the decease in the number of informal sector workers. The decline in demand for U.I.S. goods and services is attributable to a general decline in real purchasing power of economies undergoing adjustment arising from attendant inflation, wage ceilings and tight monetary policies adopted. A reduction in the capital capacity of the U. I. S. workers is due to the increased costs of household consumable goods and services which costs are offset by using funds from the working capital.

Although, 59.2% of the U.I.S. workers reported working for over eight hours a day, majority respondents maintained that the long hours of work were no different from to those put in during the pre-adjustment era. Hence, in the U.I.S, adjustment has had no effect on the number of hours put in by the operators, for traditionally, industrial sector. workers work for very long hours.

HOUSEHOLDS' ACCESS TO The INPUTS REQUIRED FOR PRODUCTION AND EXCHANGE

Land is probably the most important prerequisite for agricultural production. There were marked differences in the modes of accessing land amongst the respondents for land tenure systems differ across Uganda. Tenure practices are a mix of customary tenure, freeholds (ordinary and adjudicated), *mailo* tenure (owners and tenants), and leaseholds provided by private owners, and, or, by government under the Public Lands Act of 1969. (World Bank 1993).

It is these differences that influence variations in modes of accessing land amongst rural farmers. Family land refers to land owned by relatives usually males (father, husband, brother) access to which is obtained by virtue of consanguinity or marital bond. Males who cultivate on family land are yet to establish own households. Same with females who are not yet married, or are single parents. For most married women, access to family land is by virtue of marriage, and childbearing. Inheriting land was quite common, testifying to the widespread prevalence of parental obligation to bequeath onto their children land upon the latter's coming to age. Only 6.1% reported acquiring land from government. These were very old men who settled in formerly uninhabited virgin land. Population increase has totally eradicated this form of acquisition. However, 17.2% had leased land from government and all were commercial farmers. Purchasing of land was a more dominant mode of accessing land, though lack of cash to purchase additional land may hinder agricultural expansion, hence compromise food security.

About half the respondents received extension advice although all the respondents reported receiving the advice rarely. As put by one respondent "they only come when they hear that one has brewed beer". Asked precisely about the form of extension advice they would like to receive, advice on improvement of yields ranked highest followed by advice on acquisition of loans and inputs - 20.2%.

Manufacturers and scrap dealing were major sources of inputs required in the manufacturing category of the urban industrial sector. However, the low level of industrialization in Kampala denied the urban industrial sector workers in Kampala opportunity to obtain industrial waste from the formal manufacturing sector. Hence the reason for a greater reliance on scrap dealers for inputs which though, makes the costs of the inputs higher. The majority in the trading category of the industrial sector obtained their merchandise from wholesalers. The Urban vendors obtained their traded-in merchandise from markets where the costs are higher. For the bulk of service workers, customers were their major source of business

On acquisition of agricultural loans, 26.3% had received loans in the preceding 6 years. Ten percent had obtained credit from Co-operative Unions. On the contrary, 15.2% had received loans from Co-operative Unions, while only 8.1% had received loans from banks. Two percent had received loans from local NG0s: First, The little prevalence of borrowing by farmers from financial institutions is due to the near absence of farmer targeted credit systems within the banks which prefer lending to businessmen whose returns are higher and quicker than farmers'. Second, the bureaucratic procedures that baffle many a peasant farmers, the social distance between the sorry-barefooted-ill clad peasant farmer and the white collar necktie banking bureaucrat which precludes a meaningful dialogue that precedes advice (Manyire 1994). Hence, a lending conducive and encouraging environment was not only non-existent, but also the existing credit systems discouraged farmers from accepting credict.

Amongst the urban workers, 20.8% reported having received loans in the preceding 6 years. However, only 1.6% reported having received loans from formal lending institutions. The rest had received credit form informal lending sources: neighbours, friends, and relatives; and traders. Even then, only 5.9% had received implicit loans from traders. Traders would give merchandise to these workers on credit and the latter would pay after selling the commodities. Therefore for the majority urban workers, informal networks of neighbourhood, kinship, and friendship were their sources of loans, but the credit obtained from these sources is usually insignificant f or investment purposes; 7. 5% used the loans for the settling health bills and meeting and meeting domestic requirements. Only 13.3% reported having used the loans for investment purposes in their trades.

With regard to forms of transactions engaged in while marketing produce, majority of the agricultural producers reported engagement in cash transactions. Only 9.1% sold on credit. Differed payment based transactions were reported by 13.1%. These were cotton farmers whose marketing had yet to be liberalised at the time these data were collected (i.e. 1993). In the U.I.S cash transactions were reported by 87.5%, 10.8% reported selling on credit, while differed payment was reported by 0.8%.

Taxation of farmers' marketed output was reported by 74.7%. The several forms of produce taxation included market dues reported by 57.6%, Sales tax by 6.1%, Co-operative Union Commission by 8.1% and Development tax, reported by 3.0%.

Liberalization of product markets in Uganda has ushered in new changes in the management of the markets, which adversely affect farmers. Market places have been privatized, and are now managed by private entrepreneurs who collect the market dues and rents. Unlike prior to privatization of the markets when they were managed by District Local Administrations, market dues have been increased to between 10% and 20% of the total farmers' sales. Market dues' collectors arbitrarily determine the amount collected from the farmers since there are no set taxation criteria. The amount taxed depends on the collectors' discretion. Consequently, all farmers were of the view that taxes were very high which greatly reduced their incomes. In the urban areas, 96.7% reported paying taxes. Market dues were paid by 73.4%. Sales tax, monthly rates and income tax were paid by fewer of the respondents, particularly those who had larger establishments. Expectedly, 64% of the urban workers reported that taxes had increased tremendously

Transportation of produce from farmers' homes to the markets was the greatest problem that affects farmers, followed by price instability. Transport problems arise from the impassable nature of rural roads and near absence of rural public transport systems. Fluctuating agricultural produce prices are characteristics of the Ugandan household agricultural systems whereby prices drop to the bottom in the first few months following harvesting, and skyrocket in the 2-3 months proceeding the next harvest. Unfortunately, because of the high need for cash, majority farmers are unable to hold until prices rise, hence usually sell the bulk of their produce at rock bottom prices. In the U.I.S, fewer customers resulting from reduced demand for the sectors' goods and services was the greatest problem affecting the U.I.S workers. Inadequate shelter (in the operating premises), and difficulties in obtaining trading licenses were equally great problems. High costs of inputs required for business were a major obstacle to 26.7% of the respondents.

Farmers were overwhelmed by problems faced in marketing agricultural produce 30.3% had no circumventing solution at all, while 16.2% of the respondents sold from their homes where certainly they are price-takers, taken advantage of by traders who having realized that the peasants did not have lasting and viable storage facilities, and were not aware of the changing market prices, paid them a price less than they really deserved. Only 9.1% had the capacity to store their produce until prices rise. For the rest of the respondents, hiring costly transport and transporting small quantities using manual labour was the alternative. Only 4.0% owned bicycles, which they used to transport produce. Amongst the urban workers, 45.9% had no solutions at all to the problems they faced in the sector. Bringing additional family members into trade, especially young children were reported by 11.7% while obtaining credit was reported by 9.9%.

RURAL PRODUCERS' AND URBAN WORKERS' RESPONSE TO PRICE AND INPUT CHANGES

The overwhelming reliance on rudimentary technology (hoe, axe, *panga*) using family labour greatly held back producers' capacity to increase output in response to higher prices. 97% relied on family labour for farming. Augmenting family labour with hired labour was reported by 31%. Even then, it appears that the hired labour used the inefficient rudimentary technology for none used ox-ploughs, while only 24.2% reportedly hired tractors.

Investment in agricultural inputs, and, or equipment in the preceding six years was equally low, with over 50% of the respondents reporting not to have invested in either. Of those who had invested in agricultural inputs/equipment, 39.4% had invested in traditional tools (hoes, and *pangas*). Only 1% had invested in fertilizers.

In the Urban areas, 50% of the respondents reported having invested in inputs and equipment in the proceeding six years. The higher proportion of urban workers who had invested in inputs and equipment compared to the rural agricultural producers is not testimony to the urban workers' greater entrepreneurship exhibited in form of higher investment therein. Rather, it is indicative of the low level of ownership of durable assets amongst low income urban workers. For example, while agriculture is practiced on land (a fixed asset), and with use of fairly longer lasting implements e.g. hoes, *pangas*, in the urban informal work, inputs and equipment include hammers, chisels, saucepans, cups, plates, etc. all of which have shorter life spans and require periodic replacement. This is confirmed by the type of inputs invested in by the informal urban workers: only 6.7% had invested in high cost longer lasting equipment, and most were manufacturing informal sector workers who had purchased grinders, and electricity powered and manual metal and carpentry tools. The rest of the informal sector workers who had invested in inputs and equipment had purchased low cost items necessary for their trade though not capable of significantly increasing output of goods and services produced and subsequently, income.

Lack of capital to purchase agricultural inputs/equipment was the major reason for not investing therein, reported by 51.6% of the respondents. With regard to the agricultural input/equipment preferences of smallholder producers, they largely preferred traditional and intermediate farming equipment rather than modern tools.

The majority respondents preferred intermediate tools which were both relatively inexpensive but time efficient.

Lack of cash and failure to obtain credit were the other reasons for failure to use the preferred inputs/equipment. However, unavailability was reported by only one-tenth of the agricultural producers indicating that the Ugandan agricultural inputs' market was functioning well.

Besides failure to invest in production enhancing equipment and inputs, agricultural respondents further intimated that there were crops they would have liked to grow but were unable. The nature of these crops include Traded (25.3%), tradable crops (30.3%), non-tradable (31.3%) In percentage terms, the reasons for the respondents' inability to grow the crops included, insufficient land 19.2, Poor soils (16.2), unaffordable inputs (38.4), unfavourable climatic conditions (7.1), market pressures (2.0), poor physical conditions (2.0), long gestation period (2.0) and lack technical knowledge. These results reiterate the central role played by non-affordability of inputs in hindering rural smallholders' capacity to increase output in response to higher prices. The central role played by non-affordability of inputs becomes even more crucial considering that the other important reasons cited for not growing the desired crops. In other words, insufficient land size and poor soils could be overcome by capital intensive agriculture that could derive from affordability and subsequent adoption of high yield varieties, fertilizers, pesticides, and both mechanical and hired labour.

CONSUMPTION STRATEGIES AND ADAPTAPTIVE BEHAVIOURS DESIGNED TO SAFEGUARD FOOD INTAKE

Considering the increased expenditures on most households' responded to the increased challenges. The respondents had fewer avenues for increasing their incomes in order to obtain the sufficient food requirements in the wake of increased prices. Consequently, 26.0% of the urban workers and 28.0% of the rural agricultural producers reported doing without some food items, and consumed less of those items they could afford to purchase. In the rural areas, only 4% reported engaging in off-farm activities while only 9.0% reported selling their labour in bid to obtain the additional income required to purchase food (table 6.18).

The high proportion of respondents (over one-quarter in each location) who reported doing without some food items and, or, consuming less quantities lends credence to the high prevalence of malnutrition in Uganda. Republic of Uganda (1990b) showed that 45% of the children aged 0-60 months had stunted growth. Stunting reflects the cumulative effect of chronic malnutrition over a number of years, and it is typically associated with poor economic conditions and the attendant inadequate food intake arising from household food insecurity, inability of the body to effectively utilized food because of disease, or both (World Bank 1993d). Even where less food items were consumed, rather than doing without them completely, reported purchases of those items were in such small quantities, which certainly did not meet the entire household's food needs. Pathetic cases of going without salt for more than two weeks seemed to be the rule rather than the exception particularly in the rural areas. Items like sugar, milk, meat and *ghee* were a "luxury" preserved for only the sick, and or, consumed on festive days like Christmas.

Table 6.18: *Strategies adopted in the wake of increased food expenditures by location*

Strategies	Urban	Rural
Sell farm produce	-	14.0
Do without some food items/consume less	26.0	28.0
Obtain credit/borrowing	5.0	8.0
Relatives assist	2.0	7.0
Increase acreage cultivated	-	15.0
Forego other essentials	-	2.0
Sell animals	-	9.0
Sell Labour	4.0	9.0
Off-farm activities	-	4.0
Increase price of farm produce/products	5.0	-
Work informal sector	1.0	2.0
Sell property	-	1.0
Use income from business	34.0	-
Not applicable	23.0	26.3
Total (%)	100	-

Total (n) 100
N.B. Total % for rural Uganda is more than 100 because of multiple responses.

Farm produce sales were practised by 14% of the rural agricultural producers suggesting that these were more at risk of depleting their food reserves thus face food insecurity before the next harvest. Further, more urban workers and rural agricultural producers resorted to short term stabilizing strategies, which merely postponed the problem without solving it. These strategies included obtaining food on credit, and, or borrowing cash. Distress strategies such as selling of animals and, or property in order to raise money to purchase food was also common in rural Uganda. Given that animals are considered security premiums in most agrarian economies, selling them to purchase food is indicative of stripping oneself of vital security in event of larger life shocks, which pushes this category into a downward spiral form which recovery is rare (Chambers, 1984).

With regard to strategies adopted in the wake of the increased prices of other domestic use commodities, rural agricultural producers had less coping strategies: 24.2 % reported that they had no strategies at all seventeen percent did without some commodities while 14.1% resorted to obtaining the commodities on credit. Only 8.1% sold produce to raise money to purchase domestic use commodities. In the urban areas, 15.8% had no strategy at all, 27.5% did without some items, and 8.6% obtained the commodities on credit.

With regard to increases in Education related expenditures, rural agricultural producers had lesser coping strategies. Considering the inefficacy of the survival strategies, 17% had withdrawn children form school due to unaffordability of pay school fees! While 16% sold produce, only 5.0% had increased the acreage cultivate and only 3% engaged in off-farm activities. As noted by world Bank (1993) and followed up by Manyire (1994b), due to the decline in the real value of government financing during the long years of economic deterioration, followed by structural adjustment, the burden of financing primary education had shifted heavily to parents, with the result of "the lower income group" either not enrolling, or withdrawing children from school when the financial burden becomes to great.

In the case of increased expenditures on health, in rural areas, 13% reported that they had no strategy at all. Expressions like "we just pray", "wait for death" are evidence of overwhelming despair. Downward spiral strategies earlier mentioned, such as selling animals and spending/selling whatever is available were reported by 10.1% and 13.1%, respectively. Manyire (1992) had earlier reported pathetic cases of people who had removed iron sheets from their houses and sold them to meet health costs. And during in-depth interviews for this study, deaths attributed to poverty (failing to pay for health costs) were commonly cited.

In the urban areas, 13.3% also had no adoptive strategy at all. Borrowing and implicit credit i.e. obtaining treatment and paying later or in instalments was practiced by 16.7%, while spending/selling whatever was available was reported by 10.8%. Implicitly, for the majority respondents, the adaptive behaviours and consumption strategies designed to safeguard food intake were inadequate.

DISCUSSION

The liberalization of prices and markets falls within the broader goal of structural adjustment programmes initiated by the National Resistance Government (NRM) since May 1987, i.e. the goal of building supply response to changing internal and international relationships. Government objectives in stimulating growth in the agricultural sector are to meet the country's food requirements, generate foreign exchange, and improve the standards of living (World Bank 1993b). Currently, the government's agricultural policy is to encourage the production of diversified food and non-food traditional and non-traditional export crops (World Bank 1993b). In prices and markets so as to facilitate improved marketing and distribution systems, and enhancing incomes through either employment in the agricultural sector, or own production.

In the context of food, government is in the process of enacting the national food security policy whose overall aim is to ensure adequate supply of quality food at all times for national consumption, agro-based industries and export (Republic of Uganda 1993b). The objectives of the policy broadly focus on increased, and diversified food crop-production, improving the living standards of the majority population, improved food marketing and distribution systems, and enhancing incomes through either employment in the agricultural sector or own production.

Evidently, the proposed food security policy for Uganda is largely tied to the liberalization of prices and markets, and the presumption is that the latter would reinforce or augment the former. Liberalization of prices and markets especially of food, though including that of non-food products is implicitly construed sufficient to ensure household food security in particular, and the national food security in general. Although government appears to assume that the process of reforming markets and market structures will improve incomes and prices (of consumers and producers, respectively) as mediated by the positive impact on factor payments, prices, employment and output thereby ensuring food security, it does not recognize the role played social differentiation of the consumers and producers with regard to access to production inputs, employment, level of incomes, and resource inflows/outflows all of which determine the extent to which price and market incentives will be felt, and consequently, influence availability and, or affordability of household food at all times, and sufficient to enable household members lead an active health life. Further, it appears that government does not appreciate the role fiscal reforms, especially reductions in the level and allocation of public expenditures on Heath and Education, play in compromising households' food security. The compromise occurs when households have to trade off meagre resources that could have been spent on food, and instead spend the resources on the households' health and education needs whose costs rise commensurate with reductions in government expenditures on those services.

The shortfalls of government conceptualization of household food security within the context of price and market liberalization in particular, and structural adjustment programmes in general notwithstanding, this study revealed that only 20% in the urban areas and 15% in the rural areas had access to enough food for household consumption at all times. And six years since the implementation of price and market liberalization! Interestingly, the food insufficiency manifested less in form of number of meals taken per day, for only 29.0% in the urban areas, and 27.2% in the rural areas reported that the number of meals taken per day since the onset of price and market liberalization had reduced. Fifty six percent of the respondents in the urban areas and 57.5% in the rural areas reported that the number of meals had taken per day were similar to those taken prior to the implementation of price and market liberalization policies. Implicitly, the food insufficiency is more in form of quality and quantity of food consumed rather than the number of meals taken per day.

Furthermore, the chapter reveals that only 13.1% of the rural agricultural producers produced all their food requirements, contrary to widely held beliefs that agricultural households in Uganda produce most of their food requirements. Consequently, there is a high expenditure on food for household consumption both in the rural and urban areas. Seventy percent of the urban respondents and 73.7% of the rural respondents reported increased expenditures on food for household consumption while 12.0% in the urban areas and 6.1% reported decreased expenditures on food for household consumption. The increase expenditures on food for household consumption had less to do with increased food quality and quantity in-take. Instead, the increased expenditures were due to increased prices of food reported by 66% in the urban areas, and 23.0% in the rural areas. Similarly 9,0% in the urban areas and 20.2% in the rural areas had decreased expenditures on food due to a decline in the incomes they earned. These findings are in agreement with those of Boyle (1998), Manyire (1994c) and Manyire (1994e) who noted that poor households that already spent form 60% to 80% of their total income on food "were forced" first to increase the proportion of food expenditure in total expenditure and secondly, to substitute cheap for expensive sources of calories.

These findings support the hypothesis that the majority of the responsents access to nutritional food necessities had declined due to adjustment's overall adverse impact on income. In the rural areas, 42,7% attributed the increased expenditures on food for household consumption to drought, which hit Uganda from late 1992 to mid 1993. It is important to note that drought most affected the commercial farmers where 13.1% are significantly attributed decreased food consumption to low production due to drought. Subsistence farmers least attributed the decreased food consumption to drought -3% while commercial-cum-subsistence farmers who were of the view that drought was responsible for decreased food consumption constituted 9.1%. Evidently, failure of rains alone does not cause food insecurity; vulnerability to drought is determined by rural poverty, trends that influence overall economic productivity, and the institutional setting within which farming is conducted (World Bank 1993c).

The institutional setting amongst the commercial farmers was not conducive to cushioning shocks like drought though resorting to non-agricultural sources of income and livelihoods. Secondly, the cropping patterns and choices of crop mix of the commercial farmers i.e. exclusive production of maize, cotton, and beans left these farmers more vulnerable to drought compared to the subsistence farmers who had a diversified cropping system: yams, cassava, potatoes, which are drought resistant, and bananas which are both perennial and more drought resistant than the annuals produced by commercial farmers. These findings therefore reject the hypothesis that there were no significant differences in accessibility to household food between producers of tradable and non-tradable products. Instead, there were significant differences, with the producers of non-tradable crops producing more food for household consumption and having more income to purchase food, than the producers of tradable crops, the commercial farmers.

Food consumption was further compromised by increased expenditures on other essential domestic use commodities, education and health, all of whose costs had increased commensurate with devaluation, inflation, liberalisation of the foreign exchange, market and, reduction of public expenditures on the social sectors - health and education. Meagre resources that could have been spent on food were instead channelled to meeting the soaring costs of these good and services.

With regard to changes in income during the preceding six years, 47.0% in the rural areas and 37.9% in the urban areas reported increases. Increased incomes were most manifest amongst the subsistence farmers compared to the commercial farmers for the commercial farmers were most affected by removal of subsidies from tractor hire services, hybrid seeds, and pesticides which they relied on much more than the subsistence farmers. Although the correlation was not significant, more males than females reported increased farm incomes, implying that males were more predisposed to feel the price incentives ushered in by price and market liberalisation than females. However for the majority rural producers, incomes had not increased with the implementation of price and market liberalisation testifying to the minimal existence of price controls of agricultural produce prior to the introduction of the liberalisation policies, with the exception of coffee and cotton.

Almost three times more males than females reported increases in urban incomes since liberalisation once again testifying to the imbalances between the sexes with regard to employment and income. These findings support our hypothesis that price and market liberalisation exacerbates more the women's accessibility to household food that men's in general, though the location of the women (urban or rural) plays a major role in determining the extent of their income earning and subsequently effective demand for food.

However, it is important to note that although respondents in the rural areas who reported increases in income were more than those in the urban areas, paradoxically, more urban respondents reported that returns from their employment (self or otherwise) met both their household food and production (investment) needs. This applies to both sexes, implying that the increases in incomes in the rural areas matched less the increases in prices of goods and services consumed compared to the increases in the incomes of urban workers. While urban males' incomes met most their household food and production needs, the incomes of urban females met their food and production needs more than those of rural males, while rural females' returns least met their food and production needs.

Therefore, urban employment cushions against vulnerability to food insecurity more than rural agricultural production which neither produces enough food to meet all the household's food needs nor is profitable enough to warrant obtaining income high enough to enable effective demand for food. These findings refute the hypothesis that the high incidence of recourse to market exchanges by urban households makes them more negatively affected by price and market liberalisation compared to rural households, majority of whom provide their own food. What is crucial is not that the degree of recourse to market exchanges but effective demand for food, which the urban workers could command more important for the rural agricultural producers.

Furthermore, these findings refute the hypothesis that consumers of non-traded foodstuffs are less adversely affected by price liberalisation than consumers of traded foodstuffs. For majority rural households consume non-tradable crops compared to a sizeable number of urban households who consume tradable crops. Besides some of the nutritious foodstuffs whose prices have increased prohibitively are non-tradable crops. These include protein foods like milk, fish and meat.

The findings of t his study did not lend support to the liberalisation theory which assumes that producers/workers increase output in response to price and market incentives, and that there is always a marked shift from the production of non-tradable crops to production of tradable crops whose prices are assumed to increase commensurate with devaluation and the liberalisation of prices and markets. Instead, 75% of the agricultural producers produced non tradable, 35.4% produced traded crops, and 51.5% produced tradable crops. Only 11.1% reported that the crops grown presently were not grown in the period preceding liberalisation. Neither had there been much increase in area cultivated. Only 10.1% reported having increased area size cultivated. Majority, 67.6% had maintained the same area sizes cultivated before liberalisation. In the urban areas, 54.1% reported a decrease in the volume of their employment activities while only 32.3% reported an increase. The decrease in volume of employment activities was attributed to the increase in the number of people joining urban employment especially in the informal sector where the respondents were engaged. Reduced capital due to channelling production funds to meet household welfare including food, needs, and reduced demand of their products and services in part due to a decline in income of the customers constituted 16.0% each.

The low output increases since the onset of price and market liberalisation were largely due to the less ability of both the rural and urban producers/workers to marshal the necessary resources required to increase output. Extension advice, credit, and investment inputs were mostly lacking in the rural agricultural sector. Taxation and poor marketing infrastructure, and reliance on manual technology for cultivation hindered the feeling of price incentives and, increasing output to the desired levels, respectively. In the urban centre taxation, high costs of inputs and lack of sufficient capital for re-investment were equally big obstacles, which impacted on incomes that in turn affected the capacity to purchase food. Amidst these obstacles 46.0% of the urban workers and 30.3% of the rural producers described themselves as having no solutions to the obstacles at all. Eleven percent of the urban workers were bringing additional family members into trade to augment the family incomes while 19.0% obtained credit. Amongst the rural agricultural producers, 16.2% sold their produce from their homes while 10.1% hired costly transport and another 10.1% ferried their produce to the markets in small quantities at a time.

The adaptive behaviour and consumption strategies designed to safeguard food intake in the wake of liberalisation were rather inadequate. 26 % of the urban workers and 28.0% of the rural producers reported that they reduce consumption, especially the expensive though nutritious food. Fourteen percent in the rural areas reported selling farm produce, which left them more vulnerable to food insecurity. And amongst the urban workers 34.0% reported using income from their enterprises, which further affected their trades.

In conclusion, the liberalisation of prices and markets enacted by the Uganda government has had varying effects on the household food security of different categories of rural household agricultural producers, and the urban workers, and between males and females. While majority urban and rural workers/producers did not have access to enough household food at all times, returns from urban employment met households' food and production needs more than rural agricultural returns. Similarly, males' returns met their households' food and production needs more than females although urban females' returns met the needs more than both the rural males' and females'.

The implementation of liberalization of the previously government controlled marketing of agricultural produce policies in Uganda did not elicit increases in incomes of the farmers because the Ugandan food economy had hitherto not been tightly controlled by government.

BIBLIOGRAPHY

Aboyade, 0. (1988) "Structural Adjustment and the African Food Economy" *Food Systems*. Harare: A Memo.

Addision, T, and Demery, L (1985) *"Macro-Economic Stabilization, Income Distribution, and Poverty: A Preliminary Survey" Working Paper 15. Overseas Development Institute.*

Adesini A.(1985) "An Economic Study of the Air Cured Tobacco Supply in Western Nigeria, 1945-64, *Nigerian Journal of Economic and Social Studies*, 12, 3.

Ardayfio, (1985) "Women and the Urban Marketing in Ghana" in R.S. Garlin and A. Spring (Eds) *Women Creating Wealth* (Washington D.C.: Association of Women in Development.

Bardan P.K. (1980). "Interlocking Factor Markets and Agrarian Development. A Review of Issues" *Oxford Economics Papers* 32 No. 1

Barduraj, K. (1975) *The Production Conditions in India Agriculture.* Cambridge: Cambridge University Press.

Bates, R. (1993) *Markets and States in Tropical Africa. The Political Basis of Agricultural Policies*, Berkeley and Los Angels: Berkeley University Press. .

Belshaw D. (1988) "Agriculture Led Recovery in Post Amin Uganda: The Causes of Failure and Bases of Success" in Hansen and Twaddle (eds), *Uganda Now: Between Decay and Development* London, Heineman Press.

Bhaduri, A. (1973) "Agricultural Backwardness Under Semi Feudalism". *Economical Journal*, 83, 1

Bunker, S. (1984) "Agricultural and Political Change in the Ugandan Economic Crisis" *American Ethnologist*, 11.

Carena, A. and Sebstad J. (1989) "Introduction: Toward a Wider Perspective on Women's Employment" in *World Development*, 17, 7.

Chambers R. (1984) *Rural Development: Putting Last First*, London, Longman.

Coclough (1988) "Zambia's Adjustment Strategy - With and without I.M.F "IDS *Bulletin* 19, 1.

Cornia et al, (1987) *Adjustment with a Human_Face*, Oxford,: Clarendon Press.

Eiswaran, M and Kotural, A (1985) "A Theory of Contractual Structure in Agriculture", *American Economic Review* 75, No. 3, June

Elson, D, (1987) "The Impact of Structural Adjustment on Women: Concepts and Issues" Paper presented at the Institute of African Alternatives Conference (London: City University).

ILO/JASPA (1981) "Employment Problems of Rural Women in Kenya Addis Ababa.", Geneval:ILO

Januakar, P.N (1989) "The Response of Peasant Farmers to Price Incentives. The Use and Misuse of Profit functions" *The Journal of Development Studies* 25, 2 January.

Jaggins, J. (1989) "How Poor Women Earn Income in Sub-Saharan Africa and What Works Against Them", *World Development*, 17, 7.

Karl, M (1 983) "Women and Rural Development : An Overview" in *Women and Development, A Resource Guide For Organization and Action* (ISIS). Geneva.

Le Franc, E, (1980) "Petty Trading in Labour Mobility: Higglers in the Kingdom Metropo1it an Area of Jamaica ", *Wor1d Development*, 17, 8.

Makgetla N. (1986) "Theoretical and Practical Implications of IMF. Conditionality in Zambia" *Journal of Modern African Studies*, 24, 3.

Manyire H. (1992) "Structural Adjustment and Agricultural Production/Productivity. The Case for small Holder subsistence and Commercial Farming" In Press.

Manyire H. (1993a) "Micro-Impacts of Economic Structural Adjustment Programmes on Traditional Support Systems and the Evolving Local Survival Strategies" A Paper presented at the *Eastern and Southern African Regional Workshop on Social Policy and Social Change in Eastern and Southern Africa: Assessing Challenges in the 1990s*, held at Colline Hotel, Mukono, 3-7, May.

Manyire H. (1993b) "Macro trends, Micro-processes, and Emerging Social Policy Issues: Some Evidence from Urban Women's Market Trading in Uganda" A paper presented at the *Annual Conference of the Canadian African Studies Association*, University of Toronto, May 12-15.

Manyire H. (1994c) "Structural Adjustment and Women in the Informal Sector: The Case for Market Women in Kampala and Jinja Towns" Research Report.

Manyire H. (1994b) "The Kibaale District Development Programme: Primary Education Assessment Study" Short-term Consultancy Commissioned". *IRISH AID*

Manyire, H. (1995): "Agricultural Intensification, the Rural Agricultural Markets and Gender: Uganda Country Study" A paper commissioned by the United Nations Research Institute for Social Development (UNRISD)'s Technical Co-operation ad Women's Lives: Integrating Gender into Development Policy Project.

Massiah J. (1986) "Women in the Caribbean" *Social and Economic Studies*, 35, 2.
Massiah J. (1989) "Women's Lives and Livelihoods: A View from the Common wealth Caribbean", *World Development*, 17, 7.
Mutemba, D. (1989) "The Impact of I.M.F-World Bank Programmes on Women and Children in Zambia" in Onimode B (Ed....) *The IMF, The World Bank, and the African Debt: The Social and Political Impact*, London, (Ed....) I.F.F.A. Co-Publications.
Nsibambi, A. (1988) "Solving Uganda's Food Problem' in Hansen, A.B. and Twaddle, M, (Eds) *Uganda Now: Between Decay and Development*, London Heinmann Press.
O'Brien, M. (1985) "The Maize Grinding Mill of Sinon" *Community Development Journal*, 20, 3.
Pinstrup-Anderson, P (1988) "The Impact of Macro-Economic Adjustment: Food Security and Nutrition". In Commander, S, (Ed....) *Structural Adjustment and Agriculture: Theory and Practice in Africa and Latin America*. London: Overseas Development Institute
Republic of Uganda (1990a) "Uganda: Accelerated Food Crop Production", Kampala, Ministry of Agriculture and Manpower for Agricultural Development Project
Republic of Uganda (1990b). "The Uganda Demographic and Health Survey 1988/89", Entebbe, Government Printers Overseas Development Institute.
Republic of Uganda (1991) "Uganda: Agricultural Sector", Memorandum "working Paper.
Republic of Uganda (1992) "Report of the 1991 Population and Housing Census". Entebbe: Government Printers.
Republic of Uganda(1993a): "Statistical Abstract 1989 - 1992 Primary Schools" Entebbe, Government Printers.
Sahn, (1990) "Fiscal and Exchange Rate Reforms in Africa: Considering the Impact Upon the Poor". CFNPP Monograph 4. Ithaca, N.Y. Cornell: Cornell Food and Nutrition Policy Program.
Sahn D. and Sarris A. (1991) "Structural Adjustment and the We l fare of Rural Small Holders; A Comparative Analysis from Sub-Saharan Africa", *The World Bank Economic Review*, 5, 2.
Sahn, D. (1 992) :"Implications of Structural Adjustment for Household Food Security in Africa" *FNA/ANA*. Vol. 2
Scobie, G. M. (1989): "Macro-Economic Adjustment and the Poor: Toward a Research Strategy" Cornell Food and Nutrition Policy Programme Monograph No.1, Ithaca, N.Y., USA, Cornell University.
Sdchultz, T. (Ed....) 1973) *Distortions of Agricultural Incentives* Bloomington, Indiana Univ. Press.
Strassman W.P(1987) "Home Based Enterprises in Cities of Developing Countries" in *Economic Development and Cultural change*, Vol. 36, No. 1
Tiragier, L(1985) "Intermediaries in the Food Marketing System of South Western Nigeria " in A. Spring and R. S. Gall in (Eds) *Women Creating Wealth*, Washington, D. C.: Association for Women in Development.
Tokman, E.(1989) "Policies for a Heterogeneous Informal Sector in Latin America", *World Development* 17, 7.
Udsholt ,L. (1994) "The World Bank and Poverty Reduction in Uganda" *CDR Working Paper No.* 94X.

UNICEF (1989) Women and Children in Uganda: A Situational Analysis. Kampala: UNICEF.

Whyte, M. A (1988) "Nyole Economic Transformation in Eastern Uganda" in Hansen and M. Twaddle (Ed.), *Uganda Now: Between Decay and Development*, London, Heinmann Press.

World Bank, (1993a) *Uganda: Social Sectors*, Washington D.C. The World Bank.

World Bank (1993b) *Uganda: Agriculture*, Washington D: The World Bank.

World Bank (1993c) *Poverty Reduction HandBook*, Washington D.C: The World Bank.

World Bank (1993d) *Uganda: Growing Out of Poverty*, Washington D. C: The World Bank.

Zuckerman (1989) "Adjustment Programmes and Social Welfare World Bank" Discussion Paper, No. 44, Washington D.C: The World Bank.

7 Structural Adjustment, Energy Crisis And Environment: A Comparative Gender Study of Rural and Urban Households

Funna Peter

Agriculture constitutes the basis of Uganda's economy. Since independence, agriculture contributed 60% of the country's GDP and 99% of its exports. Agriculture is estimated to support more than 90% of the population, mostly in producing food for subsistence or internal trade. Uganda is one of the few African nations that is self-sufficient in food, despite the almost complete collapse of its economic infrastructure due to the civil war. The country has a rich and varied agriculture. Over three-quarters of the land area receives at least the minimum rainfall necessary for intensive cultivation. 42% of the total land area of 24,341,100 hectares (60,146,858 acres) is suitable for agricultural use. However, only about 21% of the land area is currently under cultivation, mostly in the three southern regions. The major impediment to the extending cultivation and herding is the prevalence of the tsetse fly. The basic unit of production is the small-scale family holding. The average size of this holding is 1.6 hectares to 2.8 hectares (4 to 5 acres) in the south and 3.2 hectares in the north. (Kurian 1992: 2009-2010).

While coffee forms the bulk of exported crop in Uganda, petroleum forms the major energy import. However, due to the falling coffee prices on the international markets and the global recession, Uganda has experienced a moving energy debt. This has forced the Ugandan government to reconsider its policy on the energy sector. Thus the broad government policy on the energy, as outlined in the Structural Adjustment Policies (SAPs 87/88) aim at reducing the energy import burden on the Balance of Payments (BoPs) by making effective use of domestic energy sources.

Nonetheless, in the short-run, these rigorous policy measures have become a source of discord. The macro-economic reforms set in motion meso-level changes that have affected production factors and markets as well as the delivery and quality of goods services. These changes have directly or indirectly impacted on the welfare of low-income households. A counteraction between macro forecasts and micro-impacts became apparent.

In Uganda, the rural and urban areas are structurally different. 69% of the population live in the rural areas, depend on agriculture, livestock and natural resources for subsistence. Most urban Ugandans depend on various forms of wage income for livelihood and purchase their basic requirements from the market. The urban populations use two major forms of energy: kerosene and electricity. In the rural areas, the major energy source is fuelwood (firewood and charcoal). Like in the case of electricity and kerosene, prior to adjustment policies, government subsidies and price control in the energy sector were minimal, but more so in the case of the

M.A.M. Salih (ed.), Local Environmental Change and Society in Africa, 2nd Edition, 159–176.

latter. Furthermore, while the common forms of energy are marketed by institutions (like Uganda Electricity Board (UEB) and Uganda based petroleum companies), most of the fuelwood is marketed by individual producers. Thus, such structural differences in mode of energy production, consumption and expenditure patterns are most likely to be influenced independently by the adjustment measures.

Compared to location-specific structural differences, adjustment policies affect gender differently. Even if the economic system (providing food, energy and other household non-durable), does not always operate against women, the SAP measures are not gender neutral as they appear. This is because male-female differences emanating from the legal and socio-economic conditions in which women and men operate. Gender remains a fundamental principle for organising the division of labour within the family and the wider socio-economic environment. Women's reproductive and care taker roles and low bargaining power within the family and the wider socio-economic environment limits their access to, and benefits from the economic restructuring.

However, it ought to be noted that women are heterogeneous, some belong to wealthy families, others are heads of households and others are yet poor and live below the poverty line. Women contribute to the development of society but the means of production and technologies available to them are rudimentary and inefficient. Also a system in which men have a dominant role and women a subordinate position serves to decrease their access to potential benefits that may accrue from adjustment, or intensify adjustments negative effects.

With the increase in goods and services price indices, the majority of the low and middle-income households cannot afford paraffin. The situation is so bad that women headed households trade food to purchase paraffin a few others do not even have this possibility. Critically, the budgetary planning in such a situation at household level is affected. With decrease in wage rates for men and increasing rate of retrenchment by urban dwellers, it is possible for men are dependent on women income, which is already constrained. Certainly, this influences power changes at household level.

The impact of the removal of subsidies on electricity on the utilisation of charcoal and fuelwood affected urban and rural households differently. The extent of the shift from electric power to charcoal and fuelwood varies. Furthermore, with increased demand for charcoal and fuelwood leading to price increase, both the rural and urban dwellers devise different copping mechanisms in the context of falling-incomes.

Therefore, this chapter investigates the impact of SAPs on the energy sector and its differential consequences on rural and urban households. It articulates the gender analysis environmental degradation and the energy sector as well as their consequences on welfare, efficiency, equity and women empowerment.

SOCIO-ECONOMIC FACTORS

Agroforestry for sustainable development and gender specific dimensions of environmental degradation have recently been recognized to offer good prospects for tackling the fuelwood problem. Pearson and Stevens (1990) noted that inter-cropping woody biomass with agricultural crops has long been a practice in many parts of Africa. Accordingly, mono-cropping as a case with forest plantations, ignores the strength of the African traditional systems, which have been developed over a long period of time to cope with the special environmental conditions faced by Africans. Agarwal (1990); Joekes (1992) noted that because women are often disproportionately reliant on natural resources, they have a relatively high awareness of the environmental problem which affects their livelihood. Oniongo (1990) noted that when women have the opportunity to improve their livelihood possibilities, they have shown willingness and ability to protect and rehabilitate the environment.

A total of 134 respondents were interviewed. All together, composed of s64 respondents answered semi-structured interviews of which 40 were from the rural and 24 from the urban areas. Seven respondents were resource managers in the rural areas, three worked for Uganda Electricity Board officials and the rest (60) were members of the focus group discussion. The socio-economic characteristics of the respondents provided vital background information to the problem under study.

The distribution of respondents by age indicates that the youngest was 16 and the oldest 65 years of age. However, as shown in Table 7.1 below, the majority (89%) of the respondents were aged between 16 and 55 years of age, while (11%) were aged 56 years and above.

Table 7.1: *Age distribution of respondents*

	Rural	%	Urban	%
16 – 25	8	20.0	10	41.7
26 - 35	12	30.0	5	20.8
36 – 45	8	20.0	4	16.7
46 – 55	7	17.5	3	12.5
56 - 65	5	12.5	2	8.3
Total (%)		100		100
Total (n)	40		24	

The results indicate that 52% of the respondents were married, 36% were single and 12% were separated from their spouses. Of the single respondents, most of them were youth who had just dropped out of school owing to inability of parents to continue paying school dues for girls. Accordingly, 48% were female-headed households.

Education is one of the important indicators of modernisation, as it can influence people's perception, attitudes and the ways people make their decision. In this respect, education referred to the highest level of formal school attained by the respondent.

The research findings reveal that education was, not, fully accessible to the rural women as indicated by 37.5% of the rural respondents who had not attained any level of education. This implies that 70% of the rural respondents had to be tied on subsistence agriculture in the rural areas for their survival. On the other hand, only 37.5% of the urban dwellers had not attained secondary education. This implies that those with low level of educational attainment would not be absorbed in the civil service sector, therefore depending on other jobs for survival. Only 10% of rural populations had gone to tertiary institutions, compared to 20. 8% of urbanite attained the same level of education. Therefore, the urban are attracted more of the professionals compared to the rural locations. This means that the professionals had been drained away from the rural areas to the advantage of the urban dwellers.

ANALYSIS AND FINDINGS

The findings have been analysed according to the following sub-themes: For communities to initiate and sustain developmental activities like sustainable exploitation of natural resources and protection of the environment, the economic output must be on an increase. Men and women must produce not only for household consumption but for the market as well. They must generate income that is adequate to enable a decent livelihood. Maintaining survival of the family as well as providing basic services like health care and education.

Given such a scenario, the household economy is affected by intra-household income and distribution of the household needs. Farming and sale of farm produce was the major source of income for the rural populations. This accounted 57.5%, followed by salary or wage earner, which, accounted for 33%. On the other hand, 54.2% of the urban residents, their source of livelihood was from salaries and wages. This was followed by business, which accounted for 33.3% of the urban residents. Nevertheless, 10.0% of rural dwellers and 8.3% of the urban dwellers depended on donations. For the case of the rural areas these were old women and mostly got remittance from relatives and their children. On the other hand, for the urban dwellers, these were housewives who were not employed in any sector and husbands had to provide everything, which was required in the household.

However, it was noted through focus group discussions that in both rural and urban areas, some of the salary earners were also engaged in agriculture and petty trade to augment their meagre income. Under such a circumstance, both the rural and urban dwellers were being affected differently under the adjusting economy, thus the distribution of income over household needs was dissimilar among different households and location.

One of the major factors that influence household income was the household size. The dependence ratio is quite high especially in the rural areas. Surprisingly, 35.0% and 33.3% of both rural and urban household respectively did not have their own children in the household. This was attributed to both people in the location to sending their children to other relatives. As residents in the rural areas sent their children to relatives in urban areas to seek school fees and employment, parents in the urban areas sent their children to boarding schools or to relatives in the villages. Although the above case had occurred, respondents from the urban areas, 54.2% had 1-5 children in the household compared 40% in the rural area who had 6-10 children. The large number of children in the rural area was an indication of wealth and an assured source of labour force for farming. However, the average of five children in urban household was an indication of maintaining a small number, which can be catered in all aspects.

The findings revealed that it has become increasingly difficult to have a large number of children in both rural and urban areas, which cannot be catered for properly. Although both the urban and rural dwellers noted that the number of children they had was an optional number, there were complaints that their livelihood had deteriorated. As such, the household heads spent most of their income on subsistence. This task mainly affected the rural women who had to provide for the whole family as a producer. In the urban areas though, it was noted that women workers earned an income, the problem of providing for the household subsistence mainly depended on the man. Urban income-earning women only provided when the husband was not around or when they really felt that the husband did not have money.

In both the rural and urban areas it was emphasised that the tasks of providing OR supplementing on subsistence lay mainly on women. But the task of meeting major costs like education and health plus rent was a man's responsibility. Nevertheless, women headed households, had no alternative apart from working hard and meet all the household requirements.

While 19.2% of t h e rural dwellers who were supposed to be at school were kept at home, all the urban dwellers who had school children had pushed them to school. 1-5 number category accounted for the 73.1% and 93.7% for the rural and urban households respectively. The above findings revealed that the average number of children at school was five. This number is significantly high for both the rural and urban man. More specifically the urban workers depend on a meagre salary. Similarly, rural men depend on the non-tradable products often fail to pay all school requirements such as tuition fees, uniform, stationery while at the same time meet other household needs.

The focus group discussions emphasised that the urban worker suffered most because school requirements and fees were very expensive compared to village schools where the fees were subsidised. Women-headed households often abandoned by their unemployed husbands are obliged to educate their children. Furthermore, it is noted that non-affordability of education is an indicator of poverty. In fact it entrenches poverty by reducing children access to the modern labour market.

Although the respondent expressed failure in providing for the education of their own children, the study revealed that some households still accept a large number of dependants. Some dependants stayed permanently or came regularly for some assistance. It is interesting to note that 40% and 34% of rural and urban households did not have dependants. This indicated a breakage in ties among relatives in the adjusting economy where everybody had to carry his or her own burden.

The study shows that 40% and 50% of the rural and urban households, respectively had 1-4 dependants. Still, the rural residents would support more dependants. Twenty per cent (20%) of the rural had 5-8 dependants compared to only 16% of the urban household. Nevertheless, the number of dependants added another burden to both the rural and urban households. The large number of dependant in the rural as emphasised by one member of the focus group discussion was an indicator of poverty and thus leading to a poor well-fare in the household; since in such a household it was not possible to cater for everybody's need. Definitely, such dependants could not be catered for in everything. In addition, a large family in rural households was viewed as a source of cheap labour.

The fieldwork also revealed that in both the rural and urban locations they received children from their relatives because in either location they could not be catered for properly. Apart from the permanent dependants in the households, the urban residents received occasional relatives who came to seek some financial assistance. That is why one participant in focus group discussion emphasize her detaste for visitors because they continued to drain even the little she had saved for the family.

The type of assistance given to the dependants included food, clothes, medical care and accommodation. The research shows that in both the rural and urban households subsistence and general welfare, which accounted 76% of the assistance rendered to dependants.

Whereas 70% of the dependants were given school fees and 60% medical care. Subsistence and general welfare accounted for the highest percentage because some of the dependants were adults who had failed either to secure jobs in any sector in the urban areas or were women who were adult enough to be married but men had not bothered to marry them, because it would add an additional burden. In the urban areas during the focus group discussion, respondents noted that young men had opted to be single because it was increasingly becoming difficult for them to be sure where would hel get the next coin to enable him survive the following day. Similarly, women at least who were able to find where to shelter themselves were not ready to get married where the husband was not sure to provide for her the next meal. Thus adjustment was increasing a gap between young men and women. Some women noted that their friends had divorced because men were not living up to their expectations, as they would not provide for the household needs.

Nonetheless, school fees and medical care services for both urban and rural producers were reported as the most income-consuming expenditures. However, the situation is more critical in the rural area because, the medical facilities were very few and at a long distance. Another reason is that the culture of cost sharing propagated to reduce government subsidise has not been properly explained and even if explained properly, may people cannot afford the cost of medicine.

In essence, most households interviewed reported that demand for manufactured goods such as sugar, soap, paraffin and food intake is decreasing. While peoples' expectations have increased, their purchase power is declining steadily. Business profits are low and fewer customers can afford to purchase them. This compounded the economic crisis. Similarly, for the rural dweller, while foodstuffs were expected to be obtained from the garden, in addition to the industrial products were also being bought from the insecure income obtained from the end of season sell of food crops like beans, maize and coffee. Such situations at household level have compounded the poor livelihood of both the rural produce and urban work, creating an energy crisis.

ENERGY USE AND ACCESS

To ensure sustainable development, the energy needs of a household must be satisfied in terms of production, supply and utilization. According to this study, the basic energy needs and sources in both rural and urban household were established as shown in Table 7.2.

Findings have much in common with earlier studies by Agarwal (1986), Cloves (1991) and Aluma (1989), that the important source of renewable energy in large use in Africa is fuelwood (charcoal and firewood). In this case, both rural and urban household the major source of energy is fuelwood. 92% of the rural and 66% of the urban households rely on firewood for cooking. Similarly, 92% of the urban households depend on charcoal for cooking. This implies that the urban households have access to both charcoal and firewood, compared to the rural households, which solely depend on firewood for cooking. In this case, most of the fuelwood is being consumed in the urban areas than the rural. Thus in terms of fuelwood needs, urban household are better off than the rural households.

Furthermore, respondents reported that electricity is abundant in both rural and urban areas. Surprisingly only 20.8% of the urban residents used electricity for both cooking and lighting compared to a mere 2.5% and 4% in the rural areas used for cooking and lighting respectively. This study revealed that the presence of electricity in an area did not mean being used.

Table 7.2: *Sources and domestic use of energy for rural and urban households*

Source	Major use	Rural (%)	Supply	Urban (%)	Supply
Firewood	Cooking	92	Moderate	66	Moderate
	Lighting	2	Moderate	0	0
Charcoal	Cooking	2.5	Scarce	92	Abundant
	Lighting	0	0	0	0
Electericity	Cooking	2.5	Abundant	28	Abundant
	Lighting	4.0	Abundant	28	Abundant
Parafin	Cooking	0	Moderate	46	Abundant
	Lighting	93	Moderate	80	Abundant
Gas	Cooking	0	Very scarce	4	Scarce
	Lighting	0	Very scarce	0	Scarce

Paraffin was reported as another source of energy. However, 93% of the rural households use paraffin mainly for lighting purposes. In the urban area 80% of the households used paraffin for lighting, and 46% used for cooking. This indicates that both rural and urban households consume reasonable quantities of the fossil non-renewable energy mostly in terms of lighting. Considering other renewable energies like biogas, solar and wind, it is embarrassing that only biogas is being used by only 4% of the urban households, for cooking purposes. In the rural households, none is using biogas, yet the production potential is more in the rural areas than in the urban.

In view of the above, urban households have more access to different energy sources for both lighting and cooking. The urban households use firewood, charcoal, electricity and paraffin for cooking. Rural households are limited to fuelwood and firewood in particular. For lighting purposes, the urban households the use of both paraffin and electricity compared to their rural counterparts who rely on paraffin. Nevertheless, 2% of the rural residents use firewood as a source of light. This is attributed to poverty (limited incomes or no income at all), because they cannot afford a smallest unit of paraffin at Shs.50=. Despite the abundance of electricity, 4% of the rural residents use electricity. These are well-to-do middleclass who are not depending on agriculture produce for their livelihood. Therefore, the energy crisis has hit more the rural dweller, though this depends not only the access but also on the social status of the individual within a particular location.

According to the focus group discussions in both the rural and urban location, energy sources for cooking we re ranked as shown in table 7.3 below. In both the rural and urban focus group discussions, participants noted that energy use and access depends on the social status of an individual. The participants accordingly divided themselves into clusters and their percentages were estimated as below:

Accordingly, the participants emphasised that these classes in society determine what type of energy to use. In this case when participants were asked to relate what type of energy accessible and used by a particular class, the following distinctions were drawn. The better off consist of the elite, business operators, women with well to do husbands and running domestic projects, these have access to electricity, gas and paraffin. These are able to even purchase the associated gadgets like stoves, cookers and refrigerators, which their counterparts cannot afford.

Table 7.3: *Energy sources by ranking*

Rural	Urban
1. Firewood	1. Charcoal
2. Charcoal	2. Parafin
3. Parafin	3. Electericity
4. Electericity	4. Firewood
5. Cow-dung	Biogas

The second category consists of those who can meet the basic needs i.e. with relatively good livelihood possibility, especially the professionals often assured of a monthly salary. Their incomes enable them to buy charcoal and firewood and appliances that go with them such charcoal and paraffin stove.

The third category identified is that of households depending, mainly on agriculture or farm produce for their livelihood. Their incomes are limited, seasonal and mostly committed to the household immediate needs. Their main source of energy is firewood and wood wastes that they collect. The cooking stoves they use are locally made.

Another category that could be considered worse off, included widows, women with very poor husbands, alcoholics, landless and casual labourers. These are poor peoples often depend on begging and remittance to eke their living. In any case, the money obtained is not enough for subsistence. Therefore, in rural areas, they use grass, dry leaves and in the absence of these, they can go hungry. In the urban, such households use briquettes, wood wastes and in the absence of they can go hungry.

FACTORS RESPONSIBLE INFLUENCING THE ENERGY CRISIS AT THE HOUSEHOLD LEVEL

Energy supply and consumption increase in the developing countries is not a question of luxury, but a requirement for the removal of poverty (Foley, 1986), because the majority of the population's basic services such as health care, education and transport require large increase in energy supplies. Accordingly, this study fuelwood is the major primary source of household energy for cooking in both urban and rural areas. Thus, it was required to assess why this was the major source.

It is interesting to note that this study's findings objected to the widely held beliefs that rural residents forage their fuelwood in their vicinity. Eleven percent (11%) of the rural residents noted that they use fuelwood because it is cheap compared to other sources of energy. Fifty percent (50%) of the urban residents attributed in creased fuelwood use to its reliability in supply compared to the other types of energy. Therefore, this means increased exploitation of the available woodlands. Contrary, this study revealed that 10% of the urban residents have planted their own trees in the peri-urban areas. In any case, if they depleted the rural sources, they can ensure are liable supply to an already established market.

The major reason advanced by the 18% rural residents for planting their own trees was that fuelwood sell had a large market in the period of scarcity, not only limited to local consumption but for export demands to neighbouring towns in the region and across the Kenya borders. This means increased exploitation of the available tree resources to satisfy external demand. However, it is not possible to satisfy the market because production is still low arid limited to the rich. Focus group discussants emphasized that markets within the rural areas but external markets in the urban areas where it fetches good money do not cause charcoal burning. For instance, in the rural areas, a sack was costing only 1,000 shilling yet in Kampala it costs as much as six folds the rural price. Therefore, the major factor for the fuelwood crisis in the rural areas is the high demand for fuelwood in urban centres. However, for the urban areas, the fuelwood crisis is attributed to the high market prices to procure the fuelwood.

While it was revealed by 75% urban residents that fuelwood was cheap and 50% because of reliable supply, 58% of the rural residents use fuelwood due to lack of substitute sources of energy. This study further revealed that keeping other factors constant, increased fuelwood use is attributed to culture values of the Ugandans, more especially where they come from. Focus group discussion participants noted that even if the prices were increasing on fuelwood, urban residents want to have a delicious meal. The aroma produced by the food, which is cooked on fuelwood is not the same as that produced by food cooked by other fuels. In any case, a woman was not ready to be beaten by a husband because she has used paraffin or electricity to cook food. One participant noted that electricity and paraffin were sources of energy for modern men and women; those people who are in a hurry and they do not care about the taste of food they prepare.

Nevertheless, fuelwood being the primary source of energy to both urban and rural dwellers, there are factors, which affected close substitute of energy for cooking. With regards to alternatives to fuelwood, the study findings revealed that 52.5% of the rural residents can do without cooking. This means that fuelwood was the major source of energy in the rural areas, therefore, its absence implies going without food, which was dangerous to the health of women and those they care for. During the focus group discussions, participants emphasised that, all the same, the firewood has to be sought least you forfeit your marriage or you are constantly harassed by the husband when he is demanding for the supper you have not cooked. Those who do not have husbands were the ones who could afford to stay without cooking, because no one would harass them.

On the other hand, 25% of the urban respondents noted that they would not cook, they had some sources of income. Therefore, they would simply visit a kiosk, have a snack for supper; which was unlikely to happen to the rural dwellers who are already constrained with their income from the sell of agricultural produce and leave alone having been cheated by the middle men.

The other major source of energy for cooking as revealed by 63.3% of the respondents was paraffin. Nevertheless, it was dominated by the urban dwellers (45.8%) compared to 17.5% rural residents who could afford buying paraffin. Apart from the increased prices of paraffin being prohibitive to the rural dwellers, they do not have access to pumping stations. Yet most of the urban dwellers have access to pump stations where they buy paraffin at an optimal price. The rural residents have to wait until the paraffin smuggled from the Kenya border is made available in weekly markets, which are in the radius of five miles away. Alternatively, they have to wait until nearby trader's travel 45 kilometres to a major urban centre to avail the paraffin at an inflated price than the pump price and the smuggled paraffin.

It was observed that in the rural areas, paraffin is not affordable because, in addition to the pump price of Shs.800=a litre, other charges like transport, tax and insurance are compounded and borne by the consumer. Meanwhile, to the urban dweller, because of the access to the pump station, the cost for fuel is half the rural price. Thus, aggravating the energy crisis in the rural areas. This was emphasized by a salesman of TOTAL (U) Ltd. , that the increased energy crisis in the rural areas as pertains to paraffin, was due to lack of market in rural areas. Also the rural dwellers especially near the boarders smuggled fuels. This affects the internal sells prices, because smuggled fuel is cheaper. Therefore, pump stations are not viable in the rural areas because of the low demand for the petroleum products. This means increased suffering for the rural women due to lack of well-developed infrastructure. Hence increased dependence on fuelwood, which is accessible.

Further, the study revealed that 35.9% would depend on crop residues as a source of energy for cooking. However, as indicated by 27.5% of the respondents, because this affected their agricultural productivity. More especially, the banana leaves which would have been used to mulch their gardens were being used as a source of household energy for cooking. In addition, the crop residues were being depleted. Thus, due to scarcity were being sold by those who had banana plantations. Not only were these leaves being used to cook food, but were helpful in the home-based industrial activities like brewing of local beer. Alternatively, cow-dung was a close substitute to crop residues, but it was limited to animals owned and the seasonality. It is only possible to use cow-dung in the dry season, the same applies to the crop residues, due to scarcity, the crop residues were also being sold to raise income for the households.

Furthermore, the study revealed that urban residents used dry leaves and twigs (crop residues) picked from the nearby trees. However, this had no significant effect on the environment and in any case they would be cleaning their environment. The alternative to crop residues to the urban dwellers, were the wood wastes. These were generated from the wood based industries in the urban areas obtained free of charge because, there was no value attached to them. This is a safety catch for the poor households in the urban areas. However, the wastes are also limited to use due to seasonality. They are convenient to use only during the dry season.

This study reveals that the hydro-electricity had spread naturally with the natural grid, but it is confined more to the urban residents. 20.8% of the urban residents used. It is a close substitute to fuelwood, though it was intermittent in supply. Despite this worthwhile picture, most of the urban residents are not accessible to electricity, as they cannot afford its use. The affordability of electricity is made difficult due to removal of government subsidies and an imposition of a tax on every consumer. Furthermore, those who have access, power supply is not constant, experiencing a load shading at least twice a week. Some few places in the urban area have special consideration, in that load shading is a myth to them. These are residential areas where the Government Ministers are staying and other privileged persons, who in any case they do not meet electricity bills personally. Therefore, those affected with constant load shading have to devise other means of cooking their food and providing light other than depending on electricity. This means increased use of fuelwood and paraffin.

In rural areas, electric cables pass over the villages towards distant urban centres. At least 4% of the rural residents have electricity in the rural town centres. The interior of the rural setting electricity is unheard of. The power lines passing over the rural households or on the rural roadsides supply 33KV to power stations, this requires a transformer, to step down the voltage to 11KV. According to the UEB officials, this could have been possible if the rural areas are potential customers who could demonstrate the propensity to consume and pay for the power supplied. In any case, they are not in position to share the cost of installing the transformer in their area, which is willingly done by the urban resident s. Thus, they have to watch the power lines pass over them after all agricultural productivity and processing needed only solar energy. Moreso, this has compounded the energy crisis on the rural dweller, as they have to depend more on the renewable energy particularly firewood, which is also very scarce. Thus affecting women and those they care for.

CONSTRAINTS FACED BY RURAL AND URBAN HOUSEHOLDS

Energy production, supply and utilization have diverse implications for Uganda's socio-economic development and environment. While growth indirectly depends on the availability of energy, the multifaceted process of energy use ranging from harvesting to end use have adverse effects on the environment (Aluma, 1989). However, at different social levels, energy production is a function of the nature of location, access to resources and the degree of overall commercialisation of the energy sector. This suggests that the household energy type is consistently mediated by a household's access to productive resources. An attempt was to study the common problems or constraints faced by the workers and rural producers, and their accessibility to factors of energy production.

Although energy is required throughout the year, seasonality and income levels of the household affect its accessibility and availability. Seasonality in the rural area dictates what activities the populace was to be engaged in. For instance, during the dry season, from January to Mid-march, off-farm activities dominate the daily tasks carried out by the individual. In this case, women are somehow relaxed because activities like land clearing and ploughing of land are mainly men's domain. Therefore, a rural woman is able to collect enough firewood and this ensures energy security for the household. However, this trend of affair is peculiar in the male-headed household. Nevertheless, in female-headed households the woman has to prepare her gardens and at the same time ensure enough firewood is available.

As regards to the rain season, where the whole family is engaged in farm activities, ranging from planting, weeding and harvesting, activities that consume a lot of women's time. In the due course, women do not have time to collect firewood and at the same time no income to purchase it.

Income from the harvests only comes out after a lapse of four months. Thus, the only alternative is to purchase fuelwood from the traders. However, the focus group participants noted that as little money from the produce is realised, firewood becomes scarce as the August rains approach. As people are busy processing the harvested crops, there is less time for women to collect the firewood even the stored firewood has been depleted. Therefore, seasonality is noted to be the major factor causing the energy crisis in the rural areas.

Urban dwellers, income is a major problem, which causes the energy crisis. Since most of the energy sources in the urban area are commercialised, lack of income to purchase it requires critical changes in household budgeting.

Focus group discussions revealed that household budgeting depends on the orientations of the national budget and whether government will subsidise certain goods and services (education, health and energy). The households are aware that their budgets are affected by the fiscal year budget. One respondent noted that when the fiscal year begins most urban households consider making expenditure changes. It affects salaries that are increasingly not sufficient to cope with the prevailing socio-economic conditions. Similarly, traders start hoarding goods or hike goods' prices.

It is also observed that the budget affects everybody. However, the constant devaluation of the Shilling and and increase of petroleum prices to reflect the international markets, affects both; the rural producer. One participant noted that some times these changes are done without consideration.

Focus group discussions in the rural areas show that kerosene prices are not be prohibitive if the households have other alternative sources of income and their produce were sold at an appropriate price.

Furthermore, the study reveals other problems faced by rural households, including land shortage, lack of capital and failure to replace the rudimentary agricultural tools.

Apart from the cost push factors; the rural producers noted environmental factors were also responsible for their crop failure. Environmental factors include the deterioration of the soil fertility as a result of over-cultivation of land to produce food for the household. The Environmental Officers observed that soil fertility deterioration is due to over-cultivation as well as absence of proper soil conservation measures such as contour ploughing, terracing and reforestation. Loss of nutrients through leaching contributes to reduced productivity and the low incomes for the rural poor.

The study further revealed that households in the urban areas, were better income managers, because their incomes are more predictable – salary, wage or profit. In the rural areas, husbands often controlled the household income. Therefore, women in both urban and rural areas do not really control their incomes. In the rural areas, men are obliged to purchase seeds, hoes, *pangas*, pesticides and other household equipment. However, some urban women are employed and have different energy needs. They control their incomes and their husbands have little say on how they spend their savings.

CHANGES IN HOUSEHOLD ACCESS TO RESOURCES

Farmers income is mainly obtained from agriculture, using low-level technology to boost production. Production on all holdings is dependant on the natural environment like adequate rain, fertile, fallow to allow it recover some fertility, was abandoned to a low crop production throughout the year. This has resulted in severe erosion on the steep slopes, which were not previously cultivated. Even where erosion is not very evident, it was acknowledged that fertility had dropped as a result of over-exploitation of land.

Focus group discussants noted that over exploitation of land is due to increased fragmentation of land. Furthermore, men who traditionally did not participate in tilling, now have separate plots in order to increase production. In the same way, men are no longer left in bed while their wives are going to cultivate. But they are waking up at t he same time. Production is geared towards non-traditional cash crops. Thus small farmland holdings are intensively inter-cropped with annual and bi-annual crops in a multi-strata home gardens, resulting in leaching especially on the mountain slopes. Furthermore, it was reported that increased intensification of home gardens in the absence of a husband meant increased work for women to till the land alone in addition to other household work.

With further land fragmentation and decreased productivity, many young have decided to abandon farming altogether and are now engaged in the more profitable cross boarder trade. Cross-boarder trade involves the smuggling of kerosene from Kenya to Uganda to be sold in weekly rotating markets. Young men move from one location to another carrying loads of kerosene jerry cans on their heads. Nonetheless, close discussions with these young men revealed that mostly rich women, who are involved in other activities, and do not have the time to move from one location to another, employ them.

The co-operatives failure to market non-traditional cash crops in the rural area has been a blessing to rural women because they are able to carry their produce to weekly markets in their respective locations. Income from the agricultural produce is not enough due to poor pricing and at times of bumper harvest a litre of kerosene costs the same amount as a bunch of banana or a tin of maize therefore, claiming most a substantial portion of the farmers' income. Thus no being able to meet all the household basic needs Furthermore, due to poverty each household member that is men women, children and other household dependant s are give responsibility to carry produce to the market. The income generated from engagement in is pooled together in order to enable the household survive. A woman respondent commented that, "putting the household income the hands of men faces a lot of vagaries, thus women often hides some away for emergencies".

With struggles for survival in the era of shrinking opportunities and scarce resources, beer parties are a uniting factor for both men and women. Beer parties have ceased being merely entertainment venues. They are potential sources of income for both men and women in the rural areas. Sustainability of the social groups hinged on providing credit facilities to members, who meet regularly and contributes a fee of 1,000/per party and 500/per visitor. Subscription charges are 2,000 per individual. Membership is open to both sexes in particular couples. Other groups exclusively cater for a particular sex. The money collected from subscription fees is given as a loan to an individual to enable him/her to purchase the ingredients for the next beer party (malt, beer palettes, water and firewood). Each person is allocated a day for, depending on the fermentation period. Every member pays for his or her pint. The money is used to pay back the loan and the profits are used to cater for individual needs.

Focus group discussions revealed that some groups are very specific in identifying the way beer party profits are used. For instance, some groups for men are helpful in incrementing men to pay graduated tax, hiring of plots, paying school fees for children, dependants and building houses for young men who have just established themselves. Meanwhile women's groups are geared towards rehabilitation of homes, more especially buy in g household goods like plates mugs, bedding as well as child dresses. These groups' are helpful in both male and female headed households, but where both male and female in the same household are involved, they have better chances of gaining in case the male does not usurp the power to control the money. It is difficult to attest to the whether women have been empowered to acquire, utilise and pay back beer party loans without the husband's interference.

The income obtained from various activities to meet household energy demands, the study revealed that firewood purchase from individuals is the most common transaction form. Individuals who had experienced wood scarcity and the anticipated high prices from firewood sale established plantations in the awake of structural adjustment. Men own most wood plantations and businesswomen own a few. Even so, this has saved women from moving long distances to collect free firewood required on a daily basis. A felled tree costs 15,000/=, an equivalent of selling fifteen (15) to thirty (30) bunches of banana. The logs are carried on heads and stored on bans. This amount provides the household energy for at least a month.

However, it is not common that households store their monthly firewood needs. In poor households, women are responsible for collecting firewood on the way home from the farm. Other women have no alternative apart from begging. In worse situations, they have to sneak into neighbours' plantation to pick a few fallen pieces of wood, with a danger of being beaten up by a ruthless owner. Poor households increasingly suffer from energy shortage, due to the commercialisation of wood, traditionally a free gift of nature.

In the rural areas, wealthy women are engaged in firewood selling, traditionally a man's. The women trade in charcoal within the region or in urban centres. Focus group discussions revealed that the men are involved in felling down trees either on public land or encroach in gazetted areas to burn charcoal. Women's task is to buy the charcoal at cheaper price and transport it to the nearest urban market for sale.

There has been a considerable change in the way urban households gain access to income including joining the informal sector either to supplement their eroded income. The attractiveness of the informal sector is exhibited in the low capital and skills required for entry. Expansion of income sources is the main reason for household engagement in the informal sector. Women are pushed into the informal sector by male relatives to reduce dependence on the meagre income. In some cases, where men felt insecure on jobs and sometimes, sympathetic with their wives over working long hours, they have to leave work early and continue with the informal sector business in the late evening hours.

Demand in the informal business was noted in the food sector. This sector requires a lot of energy for its survival. The main energy source of the informal sector is fuelwood (charcoal and firewood). Women engaged in the food sector, prepare it everyday with the exception of weekends and public holidays to provide for the urban workers who cannot afford meals in restaurants and hotels. However, a decline in the demand for cooked food is attributed to a general decline in real purchasing power high inflation and the tight monetary policies adopted. A reduction in the capital capacity of the food sellers is due to consumers' failure to pay their debts in time and increased costs of household consumer goods and services. Depleting the working capital is a desperate measure often used to offset unforeseen costs. The only redeemer is the husband who contributes part of his salary to boost the business because it would be his next source of income in case he is retrenched in the next exercise.

Although, most urban workers have encouraged their wives and female relatives to be engaged in business, men began to run separate businesses, as a security against job loss. Therefore, it was difficult to find out a clear demarcation of where the formal sector ends and where the informal sector begins. In short, the structural adjustment polices have in some respect, made women more aware of the opportunities offered by the market, thus empowering them to exploit men's labour. However, the meagre benefits that have accrued to women have not succeeded in overcoming the fuel crisis, through for instance and improvement of the household income.

STRATEGIES AND ADAPTIVE BEHAVIOUR EMPLOYED TO SAFEGUARD ENERGY INTAKE

In the preceding sections, it is noted that fuelwood (charcoal and firewood) is the major source of energy both for the rural producer and urban worker. The majority of wood based consumers are those in "poverty trap" (Foley, 1986; and Okeefe et al 1986). Consequently, consumption among countries where fuelwood is an important source of energy, is the relative availability of wood (Fluerent et al, 1978). Locational availability leads to variation in consumption within countries (Hosier,1984). This study revealed that commercialisation of the energy sector is responsible for the energy crisis. Therefore, consumption strategies and adoptive behaviour are aimed at reducing household energy expenditure.

Both the rural and urban dwellers devised almost similar strategies so as to reduce expenditure on energy. Thus, reported without doing with some energy sources. The rural producer mainly depends on firewood and consuming less of the energy they can afford. The urban worker has resorted to inferior energies like charcoal because of the high electric tariff charges. Electricity is only used for lighting, ironing, boiling drinking water and operating a television set and radio set. High electric consuming appliances like cookers have been abandoned. Even bathing is limited to cold water. Furthermore, it was reported that some households could not do without some electric appliances therefore the only way to reduce expenses on electricity is to shunt the source, thus paying less the actual amount required. Shunting electric source required devotion and commitment to rectify the situation every morning and evening with a danger of being caught in the act.

Reducing energy consumption involves alternating energy sources to prepare particular meals. It was reported that some households purchase charcoal to prepare lunch and supper and leave them on the stove to steam, instead of using electricity. Paraffin is used to prepare morning tea, frying vegetables and occasionally, cooking when a meal has to be prepared at a short notice. Nevertheless, poor households reported buying charcoal in less quantities at Shl. 200 to enable them prepare a meal for a day and in case as they were staying alone they eat a single meal or prepare meals which do not consume a lot of energy.

Obtaining fuel on credit or borrowing cash is more common among urban workers than rural producers. Fuel sellers in the urban areas are sympathetic to the workers they are aware of delayed salaries. However, the debited must be a daily customer who is not likely to be labelled a bad a debtor. Nevertheless, the fuel sellers do not offer much fuel. The fuel given should is sufficient to cook meals for a single day. Therefore, incidences of urban women failing to cook because of lack of fuel are common..

Use of income from business ranked third in both rural and urban areas but these were a few individuals and in periods of money scarcity, there is a danger of using part of t h e capital. Both rural producer and urban worker reported sacrificing and foregoing other essentials and saving little money to purchase fuel. However, saving for fuel is a common phenomenon among the workers because they have a predictable source of income - unlike their rural counterparts who cannot even borrow.

Both farmers and urban worker reported occasional sale of property. Among the rural producers, sale of property referred to sale of livestock to secure income for an eventuality. Livestock is regarded as an asset and in most cases, a buffer for an eventuality. Similarly, sale of household property among the urban poor was to raise an income, which would be used to overcome a major problem, more especially to reduce tension with an impatient landlord. However, livestock is not sold particularly to purchase fuel, but once such a sale occurs some of the income finds itself to fuel.

Dependence on the natural environment involves the use of wood briquettes, where women in the rural areas have to search for a tree, which has been felled and picked the wood residues. In the urban areas, women sent their children with exception those who have to collect wood wastes at carpentry workshops.

Respondents from the rural areas reported use of cassava and wet eucalyptus stems, which are seasoned in the morning to enable them to prepare supper. Use of folded and compacted banana leaves is very common among brewers and distillers of potent gin so as to increase income for the household. Nevertheless, all these inferior fuels in the rural areas were commercialised. Thus to enable someone to cook, a bundle of banana leaves or fibres requires an equivalent of a bunch of banana which will not be produced in the long run.

Tree planting would have been the most viable strategy to solve the energy crisis in both the rural and urban location. However, in both rural and urban areas it was ranked as the least possible response. In the rural areas, the people who had taken up tree planting activity had one hand in the urban area and another in the rural setting. Respondents reported that tree planting is not possible because of excessive land fragmentation. In any case, most holders had small plots. The small unit was for crop production. Furthermore, some were landless depending on borrowed land. Therefore, tree planting was regarded a rich persons' activity or people who do not depend on land for their livelihood. Similarly, plantation owners are profit seekers interested in large-scale sales to Uganda Electricity Board (UEB), schools and hospitals. What the poor benefit are only the branches and twigs, which can be obtained free. However, if you wanted to buy a tree, you have to purchase it at the same cost as the institutions. Therefore, those who did not have the means to purchase the trees, especially men who are being harassed by their wives have to steal the trees at night or when it is raining. Women are allowed to visit the wood lot to pick leaves, which they combine with wet eucalyptus pieces to enable them to cook a meal. Tree planting and providing firewood in an adjusting economy is generally a man's job.

CONCLUSION

The chapter analysed the impact of structural adjustment programmes on the household energy among the rural and urban dwellers in Uganda. Furthermore, it explained the adaptive strategies of both men and women and their effect on the environment. Subsequently, it examined the factors leading to the energy crisis in the urban and rural areas and the circumstance under which men and women operate as well as the constraints confronting them and how they viewed their problems and respective options.

The structural adjustment policies on the energy sector enacted by the Ugandan government have had varying effects amongst different categories of rural agricultural producer households and the urban workers. While other categories who gained from adjustment were able to meet their energy needs, others were experiencing moving difficulties. For instance, some categories of urban workers experienced increments in income, others experienced decreased in income because of retrenchment thus being unable to purchase commercialised fuel. The liberalisation of the marketing of agricultural produce in the rural areas resulted in a string of middlemen engaged in marketing non-traditional agricultural cash crops making huge profits by exploiting the farmers. Low prices have adversely affected the farmers standard of living by keeping them in perpetual poverty, therefore, not being able to purchase the commercialised fuels and those which were previously free before adjustment.

The removal of subsidies on electricity and the imposition of consumer tax by government to raise revenue have negatively affected the majority of the urban dwellers' access to this service. A few of the agricultural producers were using electricity but majority has no access to it, thus limited to fuelwood. With constant load shedding and increasing tariff changes, consequently forced the urban workers abandoning electricity use and switched to fuelwood (charcoal and firewood). Hence, forcing large quantities of fuel wood to flow from rural areas to urban centres to meet the demand, which affected the rural environment and their agricultural productivity.

Although the urban residents were buying a highly priced fuelwood to meet the traders, taxes (licence and market dues) and profits, fuelwood remained cheaper compared to other sources of energy. Hence the burden of providing energy by men in household in the urban areas was less severe than the burden to women in the rural households.

Nevertheless, due to the structural adjustment's adverse impact on income for the majority of Ugandans, most people's access to different energy sources had declined. The gains brought about by structural adjustment were quickly engulfed by increased costs of goods and services arising from the implementation of these policies. On the other hand, the capacity to respond to incentives and produce more for the market amongst the rural producers and urban workers are low.

REFERENCES

Agarwal B. (1990) *The Gender and Environment Debate: Lessons from India.*

Agarwal B. (1986): Cold Hearths and Barren Slopes. *The Wood Fuel Crisis in the Third World*, Zed Books, London U.K.

Aluma J.R.W.(1989) "Uganda's Energy Crisis: A Case Study of Fuelwood and Charcoal Consumption". A paper presented to the *1st Uganda Forestry Association National Conference* held from 3rd-6th July, 1989 at Makerere University).

FAO (1981) Map of the 'Fuelwood Situation in Developing Countries. Explanatory note FAO, Rome.

Fleuret, Patrick, C. and Annet Fluerent (1978): "Fuelwood Use in a Peasant Community: A Tanzanian Case Study". *The Journal of Developing Areas*. 12 April.

Foley, G. (1986) "Fuelwood : The Energy Crisis of the Poor". *The Courier Magazine* No.95, EEC-ACP, Brussels.

French, David (1979) *The Economics of Renewable Energy Systems for Developing Countries*. USAID, Washington D.C. January.

Ghai D. and Cynthia H. (1990): "The Crisis of the 1980s in Sub-Saharan Africa, Latin America and the Caribbean: Economic Impact, Social Change and Political Implications" *Journal of World Development*, Vol.21 389-426.

Hosier, R. (1986) *Energy and Development in Africa. A Review in Wood-Based Energy for Development*. Proceedings of a National Seminar held in Dar-es-Salaam. Tanzania 2628 March, 1984.

Howe, Michael (1979): *Indigenous Technical Knowledge: Analysis, Implication and Issues*, IDS Bulletin, U.K. Vol.10 No.2. January.

Jeffrey H. L. (1989):"The Feminization of Poverty" Environment and the Poor".

Kizito Sebana (1993) *Weekly Topic*, July 16.

Kurian, G. T. (1992) *Encyclopaedia of the Third World*, fourth edition, volume III, Facts on File: New York, N.Y., pp. 2009-2010.

Lynne Brydon and Sylivia Chant (1989) *Women in the Third World: Gender Issues in Rural and Urban Areas*. Edward Elgar Publishing Limited England.

Manyire, *H*. (1992) *The Impact of SAP on Agriculture Productivity in Uganda*. Kampala: Makerere Institute of Social Research.

McCall and Skutch M. (1987) *SADC Energy Development Fuelwood Study. Malawi Country Report*, ETC Leusden.

Messiah J. (1989) "Women in the Caribbean". In *Social and Economic Studies*, 35,2.

Mwambu, G. (1992) *Rngrgy_ Strategies in Uganda*. Energy Issues and Proposals. July Issue.

O'Keefe, P. Juma, C. and Aitken J.M. (1986) "Wood in Towns: No Money to Burn", *The Courier Magazine* No.95. EEC-ACP Brussels.

O'Keefe, P. and Munslow B. (1990) "Understanding Fuelwood". A Critique of existing Interventions in South Africa. In *Unasylva* Vol.41 No.60.

Oniong'a R.K. (1992) *An Overview of Some Environmental Concerns Affecting Women in Embu District Kenya* (Geneva).

Shanahan Y. (1986) "Fuelwood and Rural Households". *The Courier Magazine* No.95. EEC-ACP Brussels.

8 Air Pollution in Addis Ababa: The Predicament of "*Chile Sefer*"

Edlam Aberra

Air pollution, ozone depletion and green house effects are no longer environmental threats that exist and affect only the advanced industrialised societies and countries. As the Salih's chapter in this volume would explain, these are intricate issues and impact disastrously on rural communities whose subsistence activities depend on the natural environment and where climate plays a crucial rule. The problem with pollutant is that they have a very long atmospheric lifetime (about 65-120 years). Globally, UNEP published already in 1992 a full report on the health and other impacts of air pollution. These included serious threats to human, animal and plant life increase in eye and skin irritation including cataracts, general suppression of the immune system, respiratory diseases etc. The case I intend to present here is not of such biblical magnitude, but its broad implications are worrisome for developing countries sense of industrial location and consequences of environmental insensitive development.

However, due to the lack of appropriate health and environmental quality control, industries are unregulated in many African countries. Environmental management in relation to industries has been relegated to the background. Due to the priority that has been given to economic development by these countries emphasis on clean production methods have been sidelined (UN 1993:50). In Ethiopia, the release of effluents by factories has until recently not been regulated. However, industries continue to release pollutants into the natural environment without any treatment, with serious health and social effects on those exposed to it.

Furthermore, air pollution has been given the least consideration as compared to land and water pollution. This is because of a common assumption by policy makers that the level of air pollution is below the danger zone, an assumption not backed by empirical evidence. The meagre literature on air pollution in Ethiopia is evidenced in the fact that even the Environmental Protection Authority and the Environmental Protection Bureau of the Addis Ababa City Government do not possess an adequate documentation of the extent of the problem. This could be explained against the backdrop of the fact that these two environmental authorities have only recently been established.

In all fairness, the Environmental Protection Bureau of the Addis Ababa City Government had conducted, in 1998, a small-scale study. The chemical processes and wastes of a number of factories, including Addis Ababa Tire Factory, were studied and part of the findings are referred to in this chapter, with acknowledgement of source of material. However, generally, there is an apparent

M.A.M. Salih (ed.), Local Environmental Change and Society in Africa, 2nd Edition, 177–202.

lack of adequate information on the sources, types, and effects of air pollution in Ethiopia. Although limited in scope, it is hoped that this chapter sheds light on some aspects of air pollution in Addis Ababa and contributes to the existing knowledge. Its findings however, can neither be localised nor generalised without qualification.

Therefore, the main objective of this study is to investigate whether the assumption that local air pollution levels in urban centres in developing countries such as Addis Ababa town are too negligible to warrant any cause for alarm. The Addis Tire Factory is used as a case study to explain the level of local environmental damage caused by this factory and its impact on the impact on lives of the inhabitants of *"Chile Sefer"* neighbourhood. Certainly, considering the interdependence of the environmental life support system (air, water and soil or nutrients), the consequences of air pollution are not localised broader relevance to the rest of Ethiopia. It is hoped that this chapter transcends *Chile Tefe*r and provokes much broader investigations of air pollution in Ethiopia both in coverage and depth.

ENVIRONMENTAL POLICY CONTEXT

In Ethiopia more than 80% of the labour force is engaged in agriculture. However, despite its importance, the agricultural sector's performance has been far from adequate. Poverty and malnutrition are endemic in most agricultural areas. To make ends meet the small scale peasant has to exploit the available natural resources. The limited and often dwindling land resources have subsequently limited any substantial increase productivity. Although aware of the environmental imperative, many peasants have been forced to ruin their environment. Inappropriate and uncontrolled land use practices are a major cause for degradation and depletion of land resources. Natural disasters, coupled with the above conditions have led to recurrent droughts and famine in many parts of Ethiopia.

To be sure, the direction of environmental activism in Ethiopia has been strongly influenced by the effect of recurrent droughts and famines (Yeraswork 1995) The issue of environmental degradation has been usually associated with agriculture. Food security being Ethiopia's main problem, priority has been given to the rehabilitation of soil and forestry resources in order to increase productivity. Issues such as soil erosion, deforestation, desertification, and overgrazing receive the most attention. In the process, the urban environment has been given much less attention than it deserves. It is within such a framework that we should understand the development of environmental awareness in Ethiopia. An environmental awareness that has understandably been dominated by land and food security issues.

Unfortunately, until 1997, Ethiopia had no environmental policy. Instead various legal instruments and proclamations concerned with environmental degradation were put in place. For instance, the penal code and civil code of 1957 and 1960 respectively contain provisions that hold individuals liable for actions causing environmental damage. Further proclamations issued in 1965 were aimed at the protection, conservation, creation, maintenance and development of forests, wildlife etc. (EPA Nov.1997:4).

A number of proclamations were made in the 1970's and 80's, which contained provisions for the protection and management of the environment. In

1984, a proclamation provided the national committee for central planning the mandate to formulate policies and issues regarding environmental concerns. In 1987, the Ethiopian Valley Development Authority was created by a proclamation which provided it with the mandate to initiate policy as well as the means for planning and implementation of environmental management programs (EPA, Nov.1997: 4).

The establishment of the Environmental Protection Authority in 1995 is a sign of the growing concern for environmental degradation in Ethiopia. In addition, the present constitution of the Ethiopian Federal Democratic Republic provides basic principles and guidelines for environmental protection and management. It is, for instance, stated in the constitution that everyone has a right to live in a healthy environment and the government will make every effort to provide such an environment.(EPA, Nov.1997:12)

One important step in setting up the legal framework for the environment in Ethiopia was the establishment of the Environmental Protection Authority by proclamation 9/1995. The authority is an autonomous body directly responsible to the council of ministers of the Federal Democratic Republic of Ethiopia. It evaluates whether the activities undertaken by various sectors are environmentally sound (EPA 1997:10). The authority has, among many others, the following duties:

I. To prepare environmental protection policy and laws; and upon approval follow up their implementation.
II. To prepare directives and systems necessary for evaluating the impact of social and economic development projects on the environment: follow-up and supervise their implementation.
III. To prepare standards that help in the protection of soil, water, and air as well as the biological systems they support and follow up their implementation.

In line with theses duties and responsibilities, EPA has prepared two draft guidelines for the industrial sector, a regulation for Environmental Impact Assessment of development projects, a framework environmental legislation as well as an environmental policy which it issued in 1997. The latter, which is the most relevant to this study, will be discussed hereafter.

The major policy framework document with respect to environmental management in Ethiopia is the " Environmental Policy of FDRE " which was approved by the council of ministers in April 1997. EPA prepared the policy in collaboration with the Ministry of Economic Development and Cooperation (MEDAC).

Urban environment of Ethiopia is vividly shown through a description of the situation in Addis Ababa. It is stated that 31% of the houses in Addis Ababa have no sanitation facilities and that there are serious deficiencies in the sewerage infrastructure as well as random defecation, which worsen the situation. Such conditions have created dangerous health and environmental problems (EPA, 1997:2). In addition to such problems, many people in our city today are victims of toxic pollutants, which are the by-products of various man made processes. It is beyond the capacity of people to control such hazardous emissions, which pose dangers to their health. Thus, it is only the government, with the help of relevant policies, laws and regulations that can take action. This policy issued by the EPA will hopefully play an important role in this respect.

The environmental policy includes overall policy goals, objectives and guiding principles, sectoral environmental policies, cross-sectoral environmental policies and institutional, legislative, monitoring and evaluation mechanisms for the implementation of the environmental policy. A thorough analysis of the policy is not within the scope of this paper. Hence we will limit our concern to what the policy says in relation to atmospheric pollution.

The third part of the policy is concerned with sectoral environmental policies. Among these 'Pollution from Industrial Waste and of Hazardous Materials' and 'Atmospheric Pollution and Climate Change' are of interest to us. Under the former there a number of policies aimed at the minimization of pollutants released from various industries. For instance the first policy issue is as follows:

> To adhere to the precautionary principle of minimizing and where possible preventing discharges of substances, biological materials or their fragments from industrial plants and personal or communal appliances or any other external sources that could be harmful, and to disallow the discharge when they are likely to hazardous (EPA 1997: 15).

Thus the environmental policy's objective is to minimise or prevent harmful discharges from industries. However, many industries continue to emit harmful substances into the atmosphere one of, which is Addis Tire Factory without any form of waste treatment being performed.

Under this section it is stated that attempts will be made" to establish clear linkages between the control of pollution and other policy areas including water resources, agriculture, human settlements, health and disaster prevention and preparedness" (EPA 1997:15). The pollution of our environment has a vast number of effects out of which the effect on human lives and health must be given priority. A policy issue such as the above could, through showing the link between human settlements and health, help reduce the impacts of pollution on humans.

It is further stated in this section that the principle of the ' polluter pays' will be introduced. The aim here, is to ensure that polluting enterprises, municipalities and *Wereda* councils provide their own appropriate pollution control facilities. If such is the case then the Addis Tyre Factory, the relevant Health Bureau and the *Wereda* (literally province) council are all responsible for attempts to reduce the effects of the emissions of ATF on the surrounding inhabitants.

The next section is deals with 'Atmospheric Pollution and Climate Change'. Here, the country is depicted as "highly sensitive to climatic variability." Hence, one of the policy issues here is " to recognise that Ethiopia's Environmental and long term economic interests and its energy prospects coincide with the need to minimise atmospheric inputs of green house gases." (EPA 1997:17).

Much remains to be done in order to successfully achieve such an objective since there are no records of the types and amounts of gases emitted by industries in Ethiopia. Due to the lack of financial and skilled manpower resources the compilation of such data will take a long time. The problem is that until the exact extent of the problem is determined damage will continue to be done. Hence, the situation calls for immediate action by EPA in terms of persuading or even forcing if necessary, industries that are likely to emit green house gases and other harmful gases to reduce their emissions.

The geographical features of Ethiopia make the situation worse. One of the policy issues under this section intends "to actively participate in protecting the ozone layer since the highlands of Ethiopia already have a thin protective atmosphere and are liable to suffer agricultural losses and adverse health effects from exposure to ultraviolet rays" (EPA, 1997:17). Here again the policy points out the importance of controlling green house gases whose effects are worsened by the topographical features of our country. With this respect, emissions from existing factories and plants that must be monitored as there is a guideline for the establishment of new projects issued by EPA which questions the environmental sustainability of any project prior to it's establishment (EPA, June 1997) but none for the existing ones.

Generally, the environmental policy does in fact provide for the control and minimization of air pollution. The question is how practical has the policy been that is, how successful has its implementation been? This question was forwarded to an interview with Tsedale Waktola at the EPA he gave the following response;.

> The policy is being discussed in the regions and we have already covered six regions. Once we finish all the regions we will call a meeting of the concerned institutions at a federal level. We will then ask their comments on the policy. Then the environmental council to which EPA is responsible must approve the policy. Finally the concerned institutions will be given copies and will thereafter be responsible for the implementation of the policy. The duty of EPA after that will be to monitor the implementation activities of these institutions.
>
> As for the necessary laws and guidelines we have drafted a framework legislation at a workshop where representatives of private businesses, NGO's and governmental organizations were present. This framework will be the basis for the other sectoral laws. Since it is a law it must be approved by the council of ministers. Then it must be translated into English. This is the most difficult task and we expect that it will take up our time.
>
> EPA has also issued some sectoral guidelines. The practicality of such guidelines is based on voluntarism. For instance, there is a sectoral guideline for industries. Thus EPA cannot yet force institutions to adhere to the guideline through legal means. EPA will have such a power only after the framework legislation is approved by the council of Ministers.

As can be seen from the statement, the EPA has prepared the necessary framework legislation. However, EPA does not yet have the legal power to take action against industries that pollute the environment. The question is 'when will it have this power?' Unless EPA gets such a power, the existence of the environmental policy or legislative framework will be meaningless. It should not, like many other policies and guidelines in our country, continue to exist only on paper and not in practice.

Table 8.1: *Wastes generated by Addis Tire Factory*

Factory Name	Hazardous Raw Materials	Wastes Generated			Quality Control Test on Wastes	Hazard-ous Product or Wastes	Existing Risk of explosion	Radiation and noise Problem	Occupational Safety Regulations	Emergence routine for Fire
		Liquid	Solid	Gascous						
Addis Tire Factory	Almost all chemicals Used in the Factory are Toxic	-Cooling Water discharge -Manual cooling cont-aining releasing agents	-Scortch Compound -Defective tires Waste -tire Scrapes from threading section	-Dust of Carbon -Gaseous Waste From the Bioler room -Talcom dust	No	Not immed-iately hazar-dous	Raw materials in use are mostly petro-chemicals, cause fire problems	There is hissing sound in the volcanising section	The factory has safety regul-ations and provides safety facility for workers. They don't use it properly	There is fire brigade in the factory. There are fire extingui-shers.

Source: The Environmental Protection Bureau of the Addis Ababa City Government

ADDIS ABABA TIRE FACTORY AND LOCAL ENVIRONMENT

The actual fieldwork was conducted in April 1999 in Saris neighbourhood, to south of Addis Ababa city. The research was not concerned with all the inhabitants of *Kebele* 57. The study was directed towards those individuals who live right behind the factory premises. As mentioned by some respondents, this particular part of the *Kebele* has been given the nickname '*"Chile Sefer"*. This term will be used to refer to the area of study throughout this paper.

The construction of the ATF started in 1970 over an expanse of 47,000-sq.m. land. The installation of production machines was completed in mid 1972 and project production was started in the same year with the help of seven Czechoslovakian and 246 Ethiopian workers. The project was expected to produce a profit of 14.5 million Birr annually. 500 stock bonds were sold out to 5 companies, which later formed the 'Ethiopian Tire Share Association' in 1971. A few years later this association changed its name into 'Addis Tire Factory Association'. Although other enterprises were incorporated under Corporation Authority in 1975, ATF has continued to be run by a board of directors.

At present, the factory is estimated to hold 70 to 80 % of the Tire market in the country. The factory has many plans to expand its production into new areas such as the production of Tires for tractors and the like. At present, the factory seems to be successful as the platinum star award for excellence and quality it won this year may show.

Addis Tire factory can be included in the category known as rubber industries. This is because its most important raw material is rubber. There are many chemical processes, which go on in the factory. However, our interest is not in the chemical processes but the by-products of these processes. Focus will particularly be on the gaseous outputs of plant.

According to Ato Seid Abdella of the Environmental Protection Bureau of the Addis Ababa City Government, the emissions of ATF that are most likely to reach the inhabitants around the factory and affect them are particulate matter and gaseous waste from the boiler.

A number of powder chemicals are used in the factory. One powder chemical that is used in large quantities is 'carbon black', which the factory imports from abroad. As this and other chemicals are weighed and added into the mills, some amount may be released into the air in the form of particulate matter because of mishandling. In order ensure the cleanliness of the internal environment of the factory such particulate matter is released into the external environment through the ducts of the factory (Environmental Protection Bureau 1998).

According to Ato Seid, smoke is released from this part of the plant. This smoke contains sulfur dioxide, nitrogen oxide as well as carbon monoxide. These are released from the combustion of what is known as furnace oil. Burning oxidizes 40-50% of the nitrogen in heavy oil so this explains the release of nitrogen dioxide (WHO, 1992: 23). Carbon dioxide which is less harmful than carbon monoxide should have been released from the process of combustion. However, the boiler is old and causes incomplete combustion that releases carbon monoxide.

In addition to the information that Ato Seid provided, table 8.1 shows the wastes generated by the ATF. The liquid, solid as well as the gaseous wastes of the factory are listed. It is also indicated that almost all chemicals used in this factory are toxic meaning that the carbon powder used by the factory is also within this category. We can also observe that the factory does not have any quality control tests on wastes. It just releases its wastes without considering how it could affect the lives of the surrounding inhabitants.

HEALTH EFFECTS OF EMISSIONS

Air pollution can have a number of effects on humans. This section will describe the effects of specific pollutants, which are released from the Addis *Tire* Factory plant. These are sulfur dioxide, nitrogen oxide, carbon monoxide and 'carbon black'

Sulfur dioxide is " a colorless odorless gas emitted ...from the combustion of coal and oil" (Elsom 1987:21). It has often been regarded as one of the traditional pollutants of urban areas. Millard describes it as "one of the most destructive pollutants in the chemical soup we breath" (1971: 21). Both man made and natural processes contribute to atmospheric sulfur dioxide. For instance, about half of the worlds sulfur dioxide emissions come from natural sources (like volcanic activity) while the other half is a result of human activity, mainly fuel combustion in coal or oil fired stations. Human-made emissions have however been increasing 4% annually (WHO 1992:16). The inhalation of air that contains sulfur dioxide in excessive amounts can have a number of adverse health effects. Some of the possible effects as indicated by WHO includes worsening respiratory sickness such as chronic bronchitis from long term exposures. Aggravations as well as causation of asthma and pulmonary emphysema have been considered to be due to sulfur dioxide (Elsom, 1987:15 and 24). Sulfur dioxide is also thought to cause irritation on the eyes and within the nasal passage ways.(Wellburn 1988:53).

Sulfur dioxide can have different effects on human health at different levels of concentration and at different levels of exposure. The effects range from worsening of bronchitic patients to paralysis and death. We cannot ascertain the exact level of concentration of sulfur dioxide in the study area because this is not within the scope of this paper. However, we can see that even at very low concentrations, sulfur dioxide can affect human health.

Oxides of nitrogen: It can be produced by natural sources such as bacterial action in the soil, volcanic eruptions etc. It is released by human activity during combustion processes at high temperatures. However, a study conducted in 1980 puts emissions of nitrogen oxides from man made sources just over half of that from natural sources (WHO: 1992:23).

Increased sensitivity to problems of the respiratory sickness can result particularly in sensitive individuals (Wellburn 1988:91). Nitrogen dioxide is described as "a respiratory irritant and long term exposure can result in irreversible lung damage. It can cause chest tightness, burning of the eyes and headaches, particularly in persons with asthmatics and bronchitis" (WHO 1992:24).

These are some of the effects that nitrogen dioxide can have on human health. We can see how the effects range from only a slight odour being detected at a low level of nitrogen dioxide in the air to possible death at very high levels of nitrogen dioxide in the air.

Carbon monoxide: Humans have been affected by Carbon monoxide ever since he learned how to break stones by heat or to burn wood with restricted flow of air to produce charcoal. It is sometimes referred to as the oldest industrial poison (Wellburn 1988:178). It is one of the most widely distributed pollutants with global emissions estimated at 2600 tons annually. This probably exceeds the combined emissions of all other major air pollutants. Emissions of CO produced by man exceed those produced from natural sources with 60% of the global CO production being attributed to man made processes (WHO 1992:26).

CO becomes harmful to humans when it combines chemically with haemoglobin in the blood. This reduces the oxygen carrying capacity of the blood forcing the heart to pump harder to supply enough oxygen for various parts of the body (Miller, 1986:29). Once again we will try to see the varying effects of CO at varying levels of concentration of CO in the air.

Carbon monoxide can have a variety of impacts on human health. When exposed to extremely high concentrations of CO death can result within a short period. On the other hand exposure to low levels of concentration can cause tiredness, and reduction in manual dexterity. Although there is a fair relationship between symptom and the level of the gas, we must take note of the fact that heat, humidity and exertion

Suspended particulate: The term suspended particulate matter refers to "a wide range of finely divided liquids dispersed into the air from combustion processes, industrial activities and natural sources" (Elsom: 1987: 19) Particulates are divided into those that are water-soluble and those, which are not. The latter consists of elemental and organic carbon, iron oxides and a variety of other chemicals (Wellburn 1988: 20). What we call organic carbon is also known as soot. It is "made up of particles strung together in long chains and it" (Millard 1971:19)

Long term exposure to particulate matter can both cause and aggravate respiratory diseases (especially asthma and pulmonary emphysema) and also damage the lungs. The organic compound present in SPM is worrying because many of them are known to be carcinogens. The young and the old particularly those with pulmonary and heart diseases, are more susceptible to SPM. Continuous exposure of young children can have serious effects as early respiratory illnesses may develop into chronic ailments later in life (WHO 1992: 20-21). In addition, particles have an indirect effect by acting as carriers for other chemicals which otherwise would not be able to get into the lungs.

CHILE SEFER

The area in which this study was conducted is found to the southern most end of the city and is known as '*Saris*'. It is an industrial area where large number of industries is located. The *Kebele* (neighbourhood) on which this study focuses is *Kebele* 57 of *Wereda* 19 which has a total population size of 17,470 with 8,329 males and 9,145 females. The total number of households in the *kebele* is 3471 (CSA, 1995). The study did cover the whole neighbourhood, but focused only on the households located right behind the factory premises. This particular part of the *kebele* has the nickname of "*Chile Sefer*".

Before arriving at *"Chile Sefer"*", one has to pass a row of well built villas, often give the wrong impression that the people in the area are well to do. One comes face to face with reality when he/she reaches the premises behind the factory. The first impression one gets is of the abject poverty that is easily visible even without entering the houses. The scene of a cluster of mud houses, numerous infants with little or no clothing running around on the street and the fences of most compounds which are made of canvas (a waste of a nearby factory). Moreover, the repulsive smell of the open swage is an indication of the health hazards and poverty under which the people live.

Another noticeable feature is that the people, both adults and children, have dirty faces, hands, feet and clothes - a sort of soothed look. Even the stones of the street and the canvas made fences are blackened. This is why it is known as *"Chile Sefer"*.

The findings of the research are presented and interpreted in the following section of this chapter. The main data were obtained from the administration of 30 questionnaires in the study area. There are four parts to the questionnaire and these include those dealing with the socio-economic background of the respondents, the social and health impacts of emissions, mitigating measures and the respondents own suggestions as to what should be done. These will be dealt with respectively hereafter. As this major source of information was supplemented by depth interviews with *Kebele* leaders, EPA officials as well as, to the extent possible, ATF representatives, information from the latter is incorporated and analysed together with the former whenever possible.

SOCIO-ECONOMIC AND LIVING CONDITIONS

This section introduces the socio-economic conditions of the respondents and how these have compelled them to live in such a polluted environment. It also relates to the rise of environmental health among the respondents. The thirty respondents were asked to indicate their gender, employment status, monthly income, educational attainment, housing status etc. in order to give the reader an insight into their socio-economic status and the conditions under which they live. In the following paragraphs, the data pertaining to such variables will be presented.

The percentage distribution of the respondents by sex was such that nearly three out of four of the respondents (73.3 %) were female. The pilot survey in the area indicated that the women were more knowledgeable about the problem since they spent most of their time at home. The males were usually at work and did not have to face the burdens of the emissions during the daytime when the factory released most of its gaseous waste. Eight males were however included in the study as a control for variations in responses due to difference of sex. In such cases, the female concerned with cooking answered questions pertaining to indoor air pollution.

Half of the respondents were employed in non-skilled manual labour. This category included factory workers, assistant drivers, and *injera* (staple food made of local grain called *teff*) bakers. A significant portion of the respondents happened to be unemployed at the time this study was undertaken. This may perhaps be explained by the fact that most of the respondents (73.3%) were females many of whom were housewives. These respondents who claimed they were unemployed were all females. Fortunately enough 80% of these had husbands on whose income they depended totally. The rest of the unemployed women (20%) said they depended for their survival on informal activities such as engaging as casual labourers.

Apart from those who were unemployed and involved in non-skilled manual labor, 13.3% of the respondents stated that they were on pension. All of these were males. Only two out of the thirty respondents (6.7%) stated that they were employed in skilled manual work. Both of these were employed in the Adei Ababa Cotton Factory situated in the same area.

The above information indicates that not even one of the respondents was employed in a professional job. All the types of jobs in which the respondents were engaged generate very low income not exceeding 500 Birr per month. The following table indicates the distribution of respondents by income category.

The majority (53.33%) of the respondents obtained an income below 151 Birr per month. The income of the rest of the respondents ranges from a minimum of 50 Birr per month to a maximum of 500 Birr. Although the maximum income was 500 Birr per month, only one respondent received such an amount. All the rest had an income of 350 or less with a total of 79.9 % having an income of less than 201 Birr. Moreover, the mean income of the respondents was only 183.6 Birr. The general trend is that all the inhabitants earn very low incomes. At a time when the purchasing power of the Birr is at its minimum it is hard to imagine how these families survive on such low incomes. The financial problem of many of the families is worsened by the fact that the average household size is large (5.93).

43.34 % of the respondents had large families with six to eleven children while only 10 % had small families with two or less children. There were twelve (40%) middle-sized families with three to five children. At the two extreme ends were found a few families, 3 (10%) with less than 2 children and 2 families (6.67%) with more than eleven children. Whatever the number of children, it would be very difficult for the households that were questioned to make ends meet with an average income of 183.3 Birr per month. As educational attainment is one of the most important factors affecting level of income the respondents were asked about their educational background.

The low income of many of the respondents is perhaps explainable by their low educational status. 5.3, 33.7 % of the respondents were completely illiterate while 13.3 % could only read and write. Out of those were enrolled in the formal educational system, 33.3 % had reached grades 1 - 6, 10.0 % had gone as far as junior high (7-8) while only 6.7 % had attained high school (9-12).

Authors on urbanization associate the in migration of the rural poor, with the consequent increase in the size and the number of squatter settlements as well as with the spontaneous expansion of cities particularly into areas unsuitable for living such as this study area (Gupta 1988: 53). With this in mind the respondents were asked to indicate their place of birth. Only 23.3 % were born in a city with a population size of 50,000 or more. All of these respondents stated that they were born and raised in Addis Ababa. The majority, (76.7 %) were from outside of Addis Ababa. From among these, 53.3 % were from totally rural areas, one person was from a rural village, and 13.3 % were from a small town while 6.7 % were from a middle-sized town. Hence, more than half of the respondents were from remote rural areas. Obviously, such migrants, with limited skills and education, have very low possibilities of getting well remunerative jobs, if any. The 23 respondents who claimed to be originally from outside of Addis Ababa were then asked how many years it had been since they came to Addis Ababa.

Out of the respondents who were immigrants, slightly more than two thirds of them (60.86%) have stayed in Addis Ababa for 21-30 years, 13.04 have stayed for 31-40 years and one of the respondents (4.34%) has stayed for less than 20 years. Most of the inhabitants have lived in *Kebele* 57 ever since they came to Addis Ababa. 83.33 of the respondents have lived in the *Kebele* for at least 21 or more years. Only 23.33 % of the respondents have lived for 10 or fewer years in the *Kebele* while 13.33 % have lived from 11 to 20 years in the *Kebele.* The average number of years lived by the respondents in Addis Ababa is 26.04, which is a long period of time. Perhaps this average was elevated by the fact that there was one immigrant respondent had lived in Addis Ababa for 57 years.

Another particular indicator of living conditions, into which the study has queried, is housing condition. To investigate into this, a number of questions were asked in relation to housing, beginning with occupancy status. Table 8.2 gives the distribution of respondents diseases reported by the espndents..

The housing crisis, also an indication of poverty, is such that, 83.3% of the respondents do not own the house in which they are living. Only five respondents (16.7%) owned the house in which they resided. Out of the former, 66.7 % had rented the house from the *Kebele,* 13.3% from an individual and one respondent claimed to be living with a roommate, that is, in a share rental arrangement.

Those who had rented the houses in which they lived in were asked how much rent they paid. The average rent was 5.78 Birr per month. However, the rent paid varied according to the type of owner of the house. For instance, the rent paid for houses owned by the *Kebele* ranged from 3 Birr to 5 Birr per month. On the other hand those who had rented their homes from individuals paid rents between 10 and 40 Birr per month in rent.

The majority of respondents lived in houses with not more than three rooms. The average number of rooms in the houses of the respondent was 2.27%. According to the findings, 16.7 % lived in a single roomed house, 40.0 % lived in a two roomed house while 43.0 % enjoyed the luxury of three rooms as compared to the rest. In the latter case the third room was usually a '*kushena**'. We have stated earlier on that the average household size is approximately six. It is hard to imagine how the families of the respondents can manage to survive within such limited space, particularly those living in one-roomed houses.

HEALTH IMPACTS OF THE EMISSIONS

One of the major objectives of this study was to find out whether the emissions of the Addis Tire Factory have negative social and health impacts on the residents of "*"Chile Sefer"*". It was not within the scope and competence of this study to conduct actual medical check-up of the respondents to find out the health effects. From the perspective of this study, what was interesting and possible was to look in to the respondent's perceptions and their families' health status and the perceived health impacts of the emissions.

The first question under this section of the questionnaire was concerned with the symptoms of respiratory sickness. The results are presented in table 8.2 below.

Table 8.2: *Distribution of respondents by reported symptoms of respiratory sickness and the frequency of occurrence*

	Frequency of Occurrence			
Symptom	Daily	Every other day	Rarely	Never
Cough	1 (3.3%)	7 (23.3%)	7 (23.3%)	15 (50.0%)
Congestion of Nose	1 (3.3%)	4 (13.3%)	1 (3.3%)	24 (80%)
Sore Throat	0 (0%)	1 (3.3%)	3 (10.0%)	26 (86.7%)
Shortness of Breath	0 (0%)	0 (0%)	1 (3.3%)	29 (96.7%)
Headaches	6 (20%)	8 (26.7%)	9 (30.0%)	7 (23.3%)
Drowsiness	5 (16.7%)	5 (16.7%)	5 (16.7%)	15 (50 0%)
Eye Irritation	2 (6.7%)	3 (10.0%)	9 (30.0%)	16 (53.3%)

Source: Field Data

Table 8.2 indicates, 50 percent of the respondents said that they have never experienced any of the symptoms of respiratory sickness other than headaches. The symptom that was the least experienced by the respondents was shortness of breath with nearly all (96.7%) saying they had never experienced it and only one (3.3%) claiming to have experienced it. On the other hand, headache was the most widely experienced symptom among the respondents. In this case, 20.0% said they experienced headaches daily, 26.7% said they had headaches every other day, 30.0% said they rarely had headaches while 23.3% said they never had headaches. Drowsiness was the next most common symptom with 16.7% of the respondents saying they felt drowsy on a daily basis. The same proportion of respondents (16.7%) said they felt drowsy every other day and rarely. 50.0% said they never felt drowsy. Next to this, coughing was most widespread among the respondents with 3.3% saying they coughed daily, 23.3% coughing every other day and another 23.3% and 50.05 coughing rarely and never respectively.

In general, what the table indicates is that the respondents did not experience the symptoms quite often except in the case of headaches. It has already been pointed out in chapter IV that carbon dioxide can cause headaches. Perhaps the high incidence rate of headaches can be explained by the fact that the emissions of the factory contain this gas chemical.

The respondents were asked if they thought the symptoms, which they said they have experienced, were a result of the emission of ATF. The responses for this question were such that 66.7 % of the respondents believed that the emissions contributed to the existence of the respiratory symptoms listed in table 8.3. One out of three of the respondents (33.3%) did not agree with this and stated that they did not think the emissions had anything to do with the existence of respiratory symptoms.

Those who believed in the relation between the emissions and the symptoms were asked to specify which symptoms exactly they thought were related to the emissions.

The symptoms thought to be related to the emissions by most of the inhabitants were cough, headaches and drowsiness. Headache was the symptom most frequently identified by the respondents as being related to the emissions of the factory with 53.3 % stating that they believed in such a casual relation. Cough came next with 30.0 % of the respondents claiming that it was related to the emissions. Drowsiness followed with 13.3 % of the respondents stating that emissions played a role in the existence of this symptom. At the other extreme end, none of the respondents thought shortness of breath was related to ATF's emissions. This was perhaps because most (96.7%) of the respondents did not experience this symptom in the first place as indicated in table 8.3. Congestion of nose and eye irritation had no relation to the emissions of ATF by 93.3% of the respondents.

Table 8.3: *Percentage distribution of respondents by agreement or disagreement with the view that the symptoms of respiratory sickness are caused by the emissions of ATF*

	Relatedness to Emissions		
Respiratory Symptom	Yes it related	No it is not related	Total
Cough	9 (30%)	21 (70%)	100%
Congestion of Nose	2 (6.7%)	28 (93.3%)	100%
Sore Throat	1 (3.3%)	29 (96.7%)	100%
Shortness of Breath	-	30 (100%)	100%
Headaches	16 (53.3%)	14 (46.7%)	100%
Drowsiness	4 (13.3%	26 (86.7%)	100%
Eye Irritation	2 (67%)	28 (93.3%)	100%

Source: Field Data

Some scholars maintain that educational attainment can affect public opinion concerning air pollution. Crenson for instance, believes that "better educated people are more likely to be bothered by air pollution than their less well educated neighbors" (1971:14). Whether a respondent believes that the symptoms are related to the emissions or not may be determined by the extent to which he/she is bothered by the emissions. Therefore, the study has attempted to look into whether there is a linkage between educational status on the one hand, and belief that symptoms of respiratory sickness are related to ATF' s emissions on the other. Table 8.4 gives the results.

Table 8.4: *Frequency and percentage distribution of respondents by educational attainment and belief in relation between symptoms of respiratory sickness and atf's emissions*

Level of education	Belief that the Symptoms of Respiratory Sickness are Related to ATF's Emissions		Total
Educational Status	Yes	No	
Illiterate	7 (63.36%)	4 (36.36%)	11 (100.0%)
Literate Only	2 (50.0%)	2 (50.0%)	4 (100.0%)
Primary(1-6)	8 (80.0%)	2 (20.0%)	10 (100.0%)
Junior High(7-8)	3 (100.0%)	0 (0%)	3 (100.0%)
High School(9-12)	0 (0%)	2 (100.0%)	2 (100%)

Source: Field Data

Table 8.4 does not reveal the existence of an association between the two variables: educational status and belief in the relatedness of symptoms of respiratory sickness to ATF's emissions. Therefore, we cannot draw the same conclusion as Crenson' s theory that educational attainment affects public opinion concerning air pollution. Out of the illiterate respondents those who believed the respiratory symptoms were related to ATF's emissions were more (63.36%) than those who did not believe in such a relation. In addition, the respondents who had higher levels of education (i.e., high school) did not believe the symptoms of respiratory sickness were related to the emissions. Thus, we cannot say that the less educated are less bothered by the emissions or vise versa.

There is also thought to be a relation between public opinion concerning air pollution and age. For instance, Crenson believes that older people (40 and above) are bothered by dirty air more than younger people (under 40) as dirty air is more of a threat to older people than to young people (Crenson,1971: 14). With this in mind, an attempt was made to relate the age of the respondents with whether or not they believed in casual relation between symptoms of respiratory sickness and ATF's emissions.

More than half of those under 40 years of age believed there was a relation between respiratory symptoms and ATF's emissions. Out of the 14 respondents who were 40 years or older, nearly two thirds said they believed there was a relation between the respiratory symptoms and ATF's emissions. In both age groups, there are more people who believe in the casual relation between ATF's emissions and the existence of respiratory symptoms. Thus, we can neither conclude that older people (40 and above) are more bothered by dirty air nor can we say the same for young people (under 40).

Apart from questions pertaining to symptoms of respiratory sickness, the respondents were asked if they actually had any respiratory sickness. If they had any such sickness they were further asked when they developed the sickness and if they had noticed any change in the last few years. For the first question, only 16.7 % of the respondents claimed to have a respiratory sickness while 83.3 % of the respondents said they did not have any respiratory sickness. One of the respondents said, "if the factory emissions caused sickness we would all have been dead. It must be that God is watching over us." Acoording to some respondents, it seems as if the emissions from the factory have not caused any respiratory sickness among the inhabitants. This may be because the emissions are having only gradual effects on the inhabitants. Such an effect has been described as a chronic effect, which becomes evident only after continuos exposure to low levels of air pollution. These are difficult to demonstrate and consequently less obvious. (Rao,1991: 79)

In addition, the few who do have respiratory sicknesses must not be ignored because these might have been caused by ATF's emissions. To find out about this the respondents were asked when they had developed their sickness and if they had seen any changes after moving into the *Kebele*. All of the respondents who had said they had a respiratory sickness (5% of the total), maintained that they had developed their sickness even before moving into the *Kebele*. As for any noticeable changes, these respondents who stated that their sickness had gotten worse in the last few years gave a unanimous negative response.

A question concerning kinds of problems if any, respondents faced on account of ATF's emissions in their day to day lives was also asked. A summary of the problems elicited and identified by the respondents are presented in table 8.5. The table shows that all the problems listed are widely experienced by the respondents. The respondents unanimously (100%) said that ATF's emission blackened the walls and furniture of their homes. 96.7% of the respondents said they could not leave food stuff to dry out in the sun because the emissions would get into it while 3.3% said they did not have such a problem. Likewise, according to 96.7 % of the respondents the emissions caused a bad smell in the area while 3.3% said they had not noticed any bad smell due to ATF's emissions. The bad smell in the area may be a result of the sulfur dioxide and nitrogen dioxide emitted by the factory as both of these are known to have odors as pointed out in the previous chapter.

Noticeably, the respondents'clothes and bodies are blackened with soot. Soot emitted by the factory did not only blacken the clothes that they wore, but also clothes that were left to dry out in the sun according to the respondents. Nine out of ten of the respondents (90 %) said they had this problem and only one out of ten (10%) claimed not to have such a problem. Two respondents stated how they would wait for Sundays when the factory is closed to wash and dry their clothes. It is perhaps because of such a problem that four fifth of the respondents (80%) of the respondents said they faced a financial problem due to excessive consumption of soap. One respondent stated the following,

> We have to constantly wash our face, hand and clothes if we want to stay clean. This is not an area in which poor people can survive because the cost of soap can go as high as two Birr per day.

Table 8. 5: *Distribution of respondents by whether or not they face other perceived problems in relation to the ATF's emissions*

	Respondents Saying:			
	Yes	%	No	%
Problem	Frequency	Percent(%)	Frequency	Percent(%)
Cannot leave food stuff to dry out in the sun	29	96.7	1	3.3
Blackening of clothes left to dry out in the sun	27	90.0	3	10.0
Blackening of walls and furniture	30	100	0	0
Excessive use of soap for washing clothes	24	80.0	6	20.0
Cough, nausea, sneezing and suffocation when the smoke comes	23	76.7	7	23.3
bad smell	29	96.7	1	3.3

Source: Field Data

The respondents complained that the emissions created the most discomfort especially during certain parts of the day. They said it would come and engulf them like a fog and one of the respondents said, " I feel like running away" when the smoke comes. Nearly all of the respondents (96.7%) claimed that they felt nauseated, suffocated, and sneezed a lot at such times. There were also others (23.3%) who claimed not to have such problems. Other problems mentioned include the fact that the emission gets into drinking water and dirty's kitchen utensils.

When asked if they had seen any significant changes in the amount of emissions being released by the factory, the following results were obtained. The results above indicate that the majority (40%) of the respondents did not notice any changes in the amount of emissions over the last few years. Those that said they had noticed a change were divided into two with 33.3% saying the situation was getting worse and 26.7% saying that they had noticed an improvement. Respondents who said the problem became worse further stated that there was particularly an increase in the bad smell after the construction of ATF's extension plant.

MITIGATING MEASURES TAKEN BY THE RESPONDENTS

We have seen how the emissions of the factory affect the respondents in many ways. The next point to be raised should be whether the respondents take precautionary measures to reduce the effect on their health and social life. A number of questions relating to this were asked in the third section of the questionnaire and these shall be discussed below.

The ATF provides daily rations of milk for its workers. This is done in order to reduce the health impact of the powdered carbon, which can have a number of adverse health effects. The respondents were asked if they took such or other measures to reduce the health effect of ATF's emissions on their health. They all stated that they did not take any measures. They were further asked what their reason for not having taken any measure was and the following responses were given in table 8.6.

Table 8.6: *Distribution of respondents by reason for not having taken measures to reduce health effect of emissions.*

Reason for not Having Taken Measures	Frequency	Percent (%)
Emissions Have Not Affected Me Enough to Necessitate Measures	5	16.7
Financial Incapability	18	60.0
Loss of Hope	7	23.3
Total	30	100.0

Source: Field Data

The table shows that 16.7 % of the respondents had not taken any measures because the emissions had not affected them enough while 23.3% said they had lost hope and did not believe they could do anything. Some of the expressions used by the latter were " What's the point if we take measures!", " What hope do we have!" and the like. The largest percentage of the respondents (60%) said they had taken any measures because they had no money to spare. This is supported by Benneh's (1996: 19) concept of financial incapability where by the poor cannot avoid escapable morbidity and mortality because of their poor economic status.

In addition, to measures to reduce the health effects of the emissions, respondents were asked if they at least do anything to stop or reduce the entry of the emissions into their house. Only one respondent said that she had taken a measure and what she did was to put plastic bags on her window to reduce the amount of dirty air that got into her house. Even then the respondent said that the emissions managed to get into the house. The rest of the respondents (96.7%) said they did nothing to stop the emissions from getting into their house. Many of them expressed the following,

> There is no point in taking any measure because we have no idea through where the dirty air gets in. Even when the doors and windows are closed it manages to get in. The only thing we know is that the dust is always to be found on our furniture.

One of the questions that the respondents were asked was about the frequency of their visits to the doctor in the previous month. 13.3% of the respondents had gone once, 3.3% said they had gone twice while 3.3% had gone three times. Four out of every five respondents (80.0%) however had not gone to the doctors at all. The reason given by the respondents for not visiting the doctors was financial incapability.

The respondents were also asked if they ever visited a doctor only to check the effect of the emissions. None of the respondents have ever done such a thing. This may be because the respondents feel that the emissions will not affect their health or because they have never been seriously sick before. If the former is their reason then this means that the respondents are not aware of the fact that the emissions could affect their health. On the other hand, the reason why the respondents do not frequently go to the doctors may be because many Ethiopians prefer to be treated at home with the use of traditional remedies. They only turn to modern medical treatment as a last resort.

ENVIRONMENTAL AWARENESS AND SUGGESTED SOLUTIONS

As this section will show, the local populations are aware of the environmental problems that confronts them, but not knowing much about industrial pollution, they are confronted with a problem in which local knowledge offers little comfort. This conclusion is reached after discussing with the respondents in a serious of questions about their responses to the health problems posed by air pollution. The first question that the respondents were asked under this section was whether the respondents were willing to move to another area if they were given the chance. All the respondents gave a unanimous response saying that they would gladly move out of the area. When asked this question, a flicker of hope was reflected on the face of all respondents. They thought that the researcher was perhaps affiliated to the government and could do something for them in this regard. Some of the expressions given here include ones like "we would be overwhelmed with happiness", "we would not think twice if we got the chance" and the like.

The respondents were asked why they had not moved out so far if they were so eager to move out of the area. Different reasons were given in response and these have been summarised in table 8.7 below.

Once again the issue of financial incapability came at the top. More than two thirds of the respondents (76.7%) said they had not moved to another area so far because they could not afford the rent in other areas. Such a response is supported by the fact that 76.0 % of the respondents pay a rent of three Birr or less. It is difficult to find other areas where the rent is so low. Thus, it would be logical to say that these respondents are forced to continue living in the study area by the simple fact that they are poor.

Table 8.7: *Frequency and percentage distribution of respondents by reasons for not having moved out of the Kebele*

Reasons for not having Moved Out of the *Kebele*	Frequency	Percentage (%)
Cannot Afford the Rent in Other Areas	23	76.7
Feeling of Attachment to the Community	1	3.3
It Would Be Unjust	2	6.7
Relative Shortage of Housing Elsewhere	4	13.3
Total	30	100

Source: Field Data

One respondent stated that the reason why she had not moved out so far was because she felt a strong feeling of attachment to the community. She said she could not bear to think of starting a new life all over again in another area. Another surprising reason given by two respondents was that they felt it would be unjust if they moved. They did not see why they should move because they had settled in the area before the factory. Lastly, 13.3% expressed their fear that if they left the area, they would not be able to find a house elsewhere owing to the serious housing shortage in Addis Ababa. The only way these respondents can move from this area is if somebody else living in another *Kebele* and paying a similar amount of rent is willing to exchange houses. However, no one would voluntarily move into " *"Chile Sefer"*" unless forced. Thus, even if the respondents had any intention of moving out of the area they couldn't because it would be very difficult, if not impossible for them to find housing elsewhere.

The respondents were then asked if they planned take measures in order to mitigate pollution problems. All respondents said they had no plans at all. Some of the expressions used by the respondents in answering this question were "what can I do", "how can we the poor fight a government institution like the factory?" and the like. Some explained how there was a lack of co-operation and solidarity among the inhabitants. This is one of the shortcomings of the respondents in that they are not acting collectively to protect their rights to live in a healthy environment. However, they can not be blamed because they feel powerless as compared to the factory, which would have a much stronger influence on the concerned bodies. After all, this would be a case of the poor and weak trying to stand against the wealthy and powerful.

The respondents kept indicating that they would like the government to do something about the problem from the moment the researcher started to interview them. They were asked to specify what they wished the government to do. Only 13.3% of the respondents suggested that government do something about the factory emissions. All the rest indicated that they would like the government to move them away from the area.

ENVIRONMENTAL ACCOUNTABILITY

Accountability refers to being held responsible for the intended and unintended effects of ones actions and omissions. The issue of whether ATF is accountable or is willing to take responsibility for its actions is questionable. The fact that ATF is not willing to take responsibility for its actions was reflected through a number of circumstances. To begin with the factory was not willing to release any information in relation to its gaseous emissions. As the researcher observed during her visit to the factory, none of the workers were willing to reveal any kind of information. After a lot of efforts and struggle against the red tape of bureaucracy, the researcher managed to secure an appointment with one of the chemical engineers. During the interview with the engineer, all questions related to the gaseous emissions that were asked by the researcher were not properly answered. In fact, the chemical engineer said the following: "We don't have any waste as such... only some carbon is released because of mishandling and not because it is a waste." He further stated that some sulfur dioxide may be release from the combustion of oil in the boiler. This, according to the engineer, was not the problem of the factory as it was the Ethiopian Quality and Standards Authority that determined as acceptable the sulfur content of the imported petroleum.

All in all, the issue of the gaseous waste of the factory was a very sensitive one to the factory officials. It seemed that every one was trying to cover up. As, the researcher was not able to get sufficient information from the factory during the first visit she made anther attempt. The researcher approached the head of the *Kebele* and asked him to request the factory to cooperate as the study was also in the interest of the *Kebele*. The factory authorities agreed but when the researcher arrived at the plant for the second time the bureaucratic red tape unrolled immediately. She was asked to bring a letter written directly to the General Manager and was told that even then she should not expect any positive response. Thus, the researcher had to rely on the little information that she obtained from the Environmental Protection Bureau of the Addis Ababa City Administration. In fact, the factory authorities were more cooperative during the first visit. All this indicates that the factory authorities were trying to delay the moment when all the lurid details of what's really going on will come out.

The material shows, the factory does not conduct any tests for its wastes. This is an indication that the factory is not willing to take responsibility for its actions. It is not willing to consider the possible consequences on the health and social life of the inhabitants of the surrounding area.

Another interesting thing is that the factory seems to be trying to shift attention from the fact that it is releasing harmful pollutants into the air. Instead of taking measures to reduce its emissions, the factory has sued some of the inhabitants. The following is the case of one of the people who have been sued by the factory. As the respondent did not wish her name to be revealed we will not mention names.

Individual Cases: The respondent from this household is an elderly woman of 45 years. She has seven children four of who are males and three are females. She is of rural origin but has lived in Addis Ababa as well as this *Kebele* for 30 years. Due to her illiteracy and lack of marketable skills, she is unemployed. Thus, she depends on her husband's income, a pensioner who receives only meagre amount of 100 Birr per month. When asked about the factory and its emissions the respondent said the following,

> We are the forgotten citizens, the government does not know us. There is no adequate clean water and no sanitation facilities. Life in this *Kebele* is unhygienic and we are highly susceptible to diseases. On top of this, the ATF releases dirty air on us-we suffers from all these problems. We are different from others who face problems of water and sanitation facilities in that we have to also with stand the smoke of the factory.

The households of this woman as well as four other households in the same compound are related to the factory in a very interesting way.

In 1984 E.C (1994 G.C) the ATF started a lawsuit against these inhabitants. The factory had a warehouse in which it stores flammable raw materials. The ATF and the compound in which the respondent lived shared the same fence. The people of the compound shared a '*kushena', which* is built against this fence. As the inhabitants use fuel wood and industrial waste a lot of smoke is released. There is a possibility of this smoke igniting the flammable materials in the warehouse of ATF, which was also situated close to this fence. Thus, the factory filed a case against the inhabitants of this compound requesting that they rebuild their '*kushenas'* not only away from the fence but on the other side of the road. This was not only inconvenient for the inhabitants but it would have also cost them money of which they have little to spare. The situation is described as follows in the words of the respondent.

> First the factory came and made its measurements and promised us that it would move us to another *Kebele*. Then it turned against our backs and sued us. We have refuse to do anything about the problem because the houses that we live in are owned by the *Kebele* - it is the responsibility of the *Kebele* to do something about the problem. On our part as long as we are paying our rent promptly we can situate our *kushena* where we like. We have also filed our complaints concerning this issue. However, since we are fighting against a big organization so the likelihood of our case being looked at objectively is very minimal.

The case is still on going and every time it is presented before a court another appointment is given. This is an ironic situation where the factory has sued the very people whose lives it is making miserable by releasing its gaseous waste. After all what are these people expected to do about the situation? If anybody can take action it is the factory itself or the *Kebele* like the respondent suggested.

As mentioned earlier, environmental policies without proactive legal instruments are of no help to the victims of environmental marginalization. The case of ATF fits well within the wider context of environmental injustice which reveals that the victims of such injustice are often part of growing impoverished and disempowered among the urban poor in the developing countries.

INSTITUTIONAL RESPONSE

This section provides a vivid illustration of the inability of the environmental institutions to intervene on behalf of those who fell victims to environmental injustice committed against them by more powerful interests. . I take these in turn:

Addis Tire Factory: The researcher cannot say much with respect to what the factory is doing about the problem because the factory was not willing to discuss this issue. However, Table 8.7 indicates that the factory doesn't conduct any tests for its wastes whether solid, liquid or gaseous. If the factory is not willing to accept the it is doing damage to the surrounding inhabitants then the likelihood of it taking measures to reduce the emissions is very low.

The Kebele: So far the *Kebele* has not done anything about the problem, although it is aware of its existence. The researcher has informally interviewed two officials at the *Kebele* about the issue and the following responses were given. The first of the two officials was Ato Thomas who said,

> The *Kebele* is well aware that the problem exists. What is surprising is that the inhabitants have never filed a case against ATF in relation to its gaseous emissions. However, there have been cases where the inhabitants have complained about the liquid effluents of the factory. At times, the factory releases liquid waste from the boiler in large amounts and this floods the area. It seems that the inhabitants have become so used to the emissions that they do not notice it that much.
>
> I myself used to work at the Addis Tire Factory. The powdered carbon used by ATF in particular causes problems both for the workers and the surrounding inhabitants. The factory is soothed with this carbon - the people, their clothes, the equipment and machinery, the asphalt in the compound and the like. The same must apply for the area around the factory. Let alone the people, even the trees just outside the factory at the back are blackened by the factory's emissions."

The *Kebele*, answer is that,

> The *Kebele* is aware of the problem that ATF is causing. We have asked the factory about this but there was no response. Once, the factory sent its engineers and told us that it wanted to move the people. However, it has done nothing until now.

The Environmental Protection Authority (EPA): The bureau has prepared and distributed questionnaires to a number if industries concerning their raw materials and wastes. One of these was ATF, which returned the questionnaire half-empty as the officials at EPA informed the researcher. The bureau has prepared an environmental policy, which deals with air pollution as described in chapter three. If this policy is effectively implemented it will help reduce the problem of air pollution in the urban areas of Ethiopia. Right now however, nothing is being done to those industries that are releasing large amounts of gaseous waste into the atmosphere.

Environmental Protection Bureau of the Addis Ababa City Administration: This bureau is more directly concerned with factories that emit pollutants than the Environmental Protection Authority. The latter is more concerned with the issuing of relevant policies, regulations etc. According to Ato Seid , a small scale study concerning the wastes of a number of factories one of which was ATF. He stated that the researchers had problems with the officials at ATF because they were reluctant to release any information. Although the bureau hopes to conduct more in depth studies, it lacks the necessary manpower and resources, as it was established only two years ago.

In short, the results above have given us an insight in to the general living conditions of the respondents. The majority of the respondents were employed in skilled manual work (50%). Many of the respondents were illiterate (36.7%) and the highest level of education reached by some of the respondents was only 10th grade. More than half of the respondents were born in remote rural areas while a total of 76.7 % were from outside of Addis Ababa. Three out of five of the respondents (60.86%) indicated that they had stayed in Addis Ababa from 21-30 years and nearly half of the respondents maintained that they had lived in Kebele 57 for the same number of years. As for housing status, the respondents did not own 83.3% of the houses and the average number of rooms per house was only 2.27%.

Concerning the health impacts, the respondents indicated that they experienced symptoms such as headaches, drowsiness and eye irritation often while other symptoms such as sore throat and shortness of breath were less frequently experienced. More than two thirds of the respondents (66.7%) believed that the symptoms of respiratory sickness were related to ATF's emissions. Other problems faced by the respondents as a result of ATF's emission included soiling of clothes, body parts and furniture with soot.

As for self-imposed measures, none of the respondents took measures to reduce the impact of the emissions on their health. The reason given for this by three out of five of the respondents (60%) was their financial incapability. All the respondents expressed their interest to move out of the *Kebele b*
but three out of four indicated that they had not moved out so far because they could not afford the rent in other areas.

The survey results further showed that more than half of the respondents (56.7%) use industrial waste as a fuel. Similarly one in two of the respondents (50%) indicated that were aware of the possible negative health impacts of cooking with industrial waste. However, almost nine in ten of the respondents (87.5%) stated that they continued to use industrial waste because of their inability to use other fuels. Some of the problems caused by the use if industrial waste as a fuel included eye irritation nausea, coughing and the like. The results also indicated that all the respondents are willing to move out of the *Kebele.*

CONCLUSIONS

Atmospheric pollution is a major problem that will soon face all nations of the world. Rapid urban and industrial growth has resulted in vast quantities of potentially harmful waste products being released into the atmosphere. The consequence has been that air pollution has and is affecting the health and well being of people, is causing widespread damage to vegetation, crops, wildlife, materials, buildings and climate.

The problem exists in both the developed and the developing countries, In fact, these days some of the rapidly expanding cities of the less-developed countries such as Mexico City, Sao Paulo, Lagos, Cairo, Beijing and Bangkok dominate the stage where problems of air pollution are voiced. The difference between these countries is that the developed countries have the financial means to combat or reduce the problem while the developing countries do not. Thus, while the level of air pollution in many developing countries is falling or at least maintained at a certain level, the level of air pollution in many third world cities in increasing at an alarming rate.

As an underdeveloped country, Ethiopia exhibits the same traits as other underdeveloped countries in terms of air pollution. The level of air pollution in the country is increasing, there is lack of awareness concerning this issue both among the polluters and the victims of the pollution, the relevant policies are not practicalised timely and so on. The worst air pollution incidents in the world have occurred in and around urban-industrial areas. Likewise, the problem of air pollution in our country is more persistent in urban areas. The growing number of industries particularly in the capital city, Addis Ababa has resulted in the release of various pollutants into the atmosphere. Unfortunately, those who are exposed to industrial wastes are in most cases the urban poor who live around industries. These people, owing to their limited financial means cannot protect themselves from the threats of the pollutants to which they are exposed.

The inhabitants of *"Chile Sefer*" live under similar conditions. This is one of the less sought after areas of the city for living in and is inhabited mainly by poor people whose average income is only 183.3 Birr per month. Some even got salaries as low as 50 Birr. The area is situated behind the compound of the Addis *Tire* Factory. The factory releases gaseous waste, a by-product of its chemical process. This emission of the factory can easily reach the inhabitants of " *"Chile Sefer"*". It major constituents include Sulfur Dioxide, Nitrogen Dioxide, Carbon Monoxide and Particulate Matter all which have negative health impacts as shown in chapter four. According to the respondents, the emission of ATF has not affected their health gravely. However the respondents may not know for sure whether the emission has affected their health or not because they have never been to the clinic to check if the emission has had an impact on their health.

In addition to having a number of negative health effects, the emission makes the life of the inhabitants weary by covering their furniture, walls of their homes, their limbs and soiling their clothes. It is a constant intruder in their daily lives and gets into anything and everything that is openly exposed.

Unfortunately, these people cannot escape from the emission of ATF. The majority of the respondents in the survey indicated that they would be happy to move out of the area. The most important reason given for why they had not done so far was because of their inability to afford the rent in other areas. This is understandable in light of the fact that the average rent paid by the respondents is only 5 Birr. This indicates that the respondents are stuck in the area because of the simple fact that they are poor.

If the inhabitants are not able to move to another area the least they can do for themselves is to take some measures to reduce the negative impact of the emission on their health. The survey results however showed that none of the respondents took any such measures. Once again, their reason was their financial incapability. The workers of ATF get a daily ration of milk to protect them from the effects of the powdered carbon that is used in large quantities. However, such a commodity is more like a luxury for the inhabitants of *"Chile Sefer"*.

In essence, this chapter has revealed that there are multiple environmental actors. Two of these actors are more prominent: The destitute inhabitants of Chile Sefer" on the one hand and ATF, on the other. The two have interests, which are in direct opposition to each other, if not irreconcilable. The inhabitants' plea for the factory to reduce its gaseous emissions while ATF is completely opposed to such a costive measure that might eat into its profit margins. Event though both actors have their own interests, there is an imbalance between the two in terms of power over the source of the disagreement - the technology with which Tyres are produced in the ATF. It is this technology that is resulting in the release of gaseous pollutants and creating grounds for confrontation between the two stakeholders. However it is ATF and not the inhabitants of "Chile Sefer" who has control over this technology. Owing to their low educational status the inhabitants do not have a clear understanding of the workings or consequences of the technology used by ATF. Even if they were aware of the consequences, they have no say over the property and the running of the Factory. In short they feel helpless.

There is however a third actor - the Government, the only institution that can intervene to mitigate the problem. The legislative, executive and judiciary bodies of the government have the power over to bring to bear on society and its diverse economic interests. Unfortunately, the study reveals that none of the concerned institutions is willing or able to take the necessary measure in a public domain, whose protection lies well within their jurisdiction powers. Worse to come is the fact that ATF is not even willing to recognise that the problem exists or that its industrial effluents harmful pollutants. All in all, the situation is one where the voices of the weaker groups are being swallowed by the more elevated and influential voices of the business elite.

REFERENCES

Benneh, George etal (1996). *Sustaining the Future: Economic, Social and Environmental Change in Sub-saharan Africa.* Tokyo: United Nations University Press.

Burns, Tom R. and Reinhard Ueberhorst (1998). Creative Democracy: Systematic Conflict and Policy Making in a World of High Science and Technology. New York: Praeger Publishers.

Couch R., Stephen and Steve Kroll Smith, eds. (July 1997). "Sociology, Environment and Social Change", in *Current Sociology*, 45, No.3.

Crenson, Mathew (1971) *The Unpolitics of Air Pollution: A Study of Decision Making in the Cities*. Baltimore: The John Hopkins Press.

Detwyler, R., Thomas et. al., (1986). *Man's Impact on the Environment.* McGraw Hill, Inc.

Ehlers M., Victor and Earnest W. Steel, 6th ed., (1986). *Municipal and Rural Sanitation*. McGraw Hill, Inc.

Elsom, Derek (1978). *Atmospheric Pollution: Causes, Effects and Control Policies*. New York, Basil Blackwell Inc.

EPA (October 1997). *Environmental Assessment Report for the Alemgena-Hossaina-Sodo Road*. Placenter Ltd.

FAO (November 1997). "Environmental Assessment Report of the Gilgel gibe Hydroelectric Project".

FAO (1997). "Environmental Policy of the FDRE".

Gupta, Avijit (1998). *Ecology and Development in the Third World.* New York, Routledge.

ILO (1992). "Environment, Employment and Development", Geneva.Markham, Adam (1998). *The Environment.* East Sussex, Wayland Publishers Ltd..

Millard, Reed (1971). *Clean Air: Clean Water for Tomorrow's World*. Newyork: J.Messner.

Miller, Tyler G.Jr (1986). *Environmental Science: An Introduction*. California: Wadsworth Publishing Company.

Rao, C.S. (1991). *Environmental Pollution Control Engineering*. New Delhi: Wiley Easter Limited.

UN (1993). *African Strategies: for the Implementation of the United Nations Conference on Environment and Development*. Addis Ababa: Economic Commission for Africa..

Wellburn, Alan (1998). *Air Pollution and Acid Rain: the Biological Impact*. United Kingdom: Longman Group UK Ltd.

WHO. (1992) *Urban Air Pollution*. Geneva: World Health Organisation..

Yeraswork Admassie (1995). *Twenty Years to Nowhere: Property Rights Management and Conservation in Ethiopia*. Uppsala: Uppsala: Department of Sociology, Uppsala University.

9 Climate Change and Local Vulnerability: Exploring the Linkages

M. A. Mohamed Salih

Some obvious linkages exist between climate change and society, including impacts on the vulnerable populations often unable to withstand long-term reoccurrence of climatic hazards such as drought. There is therefore the association between alteration of normal climatic conditions under which peasants cultivate food and pastoralists raise animals perceived by them as a way leading a healthy productive life. This situation has been approximated in global scientific findings which illustrate that climate change will certainly impact on agricultural production as a result of the shift in cropping patterns, with the probable increase of agricultural pests and animal diseases. Such a change would entail an apparent demand for irrigated agriculture, including the undesirable prospect of wetland reclamation. For local communities such as hunters, gatherers, peasants and pastoralists, the impacts of such climate change go far beyond affecting the domesticated animals and plants, wildlife and a whole range of food varieties on which people depend for a substantial part of their diet. It affects benign (produced by the household) and contingent (on others sources of income trade, tourism etc.) household food security. Closely linked to this is the increasing frequency of forest fires, loss of hunting grounds for poor often malnourished hunter gatherer who often depend on the forests for medicine and other household needs such as building material, work tools etc. Various authors explain the impacts of climate change on human health as a result of pollution deteriorating food security leading to malnutrition and disease and shifts of vector-borne diseases that could invade areas with significantly altered temperature -malaria, schistomiasis and others.

In the particular case of Africa, studies on the consequences of global climate change on local communities are scarce. However, Okoth-Ogenda and Ojwang (1995) have succinctly reported Africa-wide recent experience with the negative consequences of climate change. [1] Although the book has not directly dealt with the consequences of climate change on local communities, it is among the most serious efforts to synthesize the policy relevance of research-and-development on this important subject. Notably, the chapters have captured a wealth of literature ranging from climate change and the environment to agriculture Greenhouse Gas (GHG) emissions, transport, global environmental governance and the like. The range of socio-economic issues the authors expounded by the contributors to Okoth-Ogenda and Ojwang volume illustrate clearly the dire consequences of global climate change on African economies and by inference society and environment. Obviously, the consequences of global climate change are demonstrably traverse sectors, sub-regions, populations and livelihood conditions.

On the whole, Okoth-Ogenda and Ojwang (1995) among others illustrate that global climate change is destined to produce global changes not all of them are

M.A.M. Salih (ed.), Local Environmental Change and Society in Africa, 2nd Edition, 203–220.

necessarily of climatic nature. In my view, these changes will manifest themselves in complex consequences in diverse patterns and magnitudes on the human conditions. Owing to their global nature, these changes are spatially multi-layered (thus traversing region, nation and locality), diverse (ranging from impaired environmental functions to food insecurity), cumulative (long-term often overweigh the short-term impacts) and integrative (not only climatic, but also social, economic and political). At least five main implications can be drawn from the above: First, during the last three decades or so, much research has rightly been directed to modelling and understanding the consequences of global climate change. Some research has also rightly gone into modelling and understanding nation or region-specific causes and potential impacts of global climate change on the biosphere.

The debate on abating and/or reversing climate change trends has often been stalled by questions pertaining to who are the national/regional users and gainers should global environmental policies to curb the causes of climate change are implemented. In some cases concerns with global, regional and national interests have blurred our perception and capacity to foresee what should be done. In others, narrow disciplinary boundaries became more assertive than the compelling need for a comprehensive understanding of how global climate change may aggravate an already desperate situation. These global processes operate not only at the global, regional and national level as abstract entities, but as aggregates of diverse local realities, communities, households and individuals that make up the globe.

Second, admittedly, locality-specific research on the consequences of climate change on household food security has not kept pace with research on the global, regional and national levels. Furthermore, local level research has been scattered, project-specific and leaning towards academic/philosophical orientation rather promoting policy dialogue or improving the quality of policy interventions. It will be grossly pretentious to insinuate that this chapter will bridge this gap in our understanding of household food security and local vulnerability. However, even when locality specific research was conducted, the regional context of household production/consumption patterns and linkages to environment, policies, institutions and markets is either narrowly defined or not fully integrated with other related processes of environmental change.

Third, local communities and household aspirations are often directed towards improving their standards of living measured by standards borrowed from and through long encounters with western consumption patterns and lifestyles. Here lies the twin dilemma of increasing aspirations for modernity leading to increasing pressures on resource thus contributing to vulnerability and an array of political and social insecurities. The question here should not be treated by reference to environmental authoritarianism, but with reference to the paradoxical situation in which improvements in peoples standards of living are equally accompanied by improvements in their environmental consciousness hence sustainable resource management. The nexus between global development trends, climate change and changes in people aspirations, attitudes and lifestyles is paradoxically problematic and cannot be understood by data extrapolation at the global or national levels isolated from local realities.

A major concern in relation to climate change is its direct linkages to food production in the developing countries as recent experiences have show.[2] Although not intended to illustrate the different nature and parameters of household, national and global food security objectives, the World Bank's (1997) matrix on dimensions of food security illustrates that researchers policy makers are overly aware of the linkages. However, the dimensions of food security offered by the World Bank are inapt in addressing household-centred food policies due to vested interests associated with the operation of global food markets. [3] In explaining the matrix, the World Bank (1997:2) argues that,

> In the short-run, reducing hunger must focus at the household level with enabling actions by nations. Globally, only adequate supplies and food aid can help. In the medium-term, the emphasis must be more at the national and international levels, focusing on reducing poverty and generating sustained economic development for all. Central to that vision are concerted national and international efforts to generate appropriate agricultural technology to improve the productivity and profitability of millions of farmers in developing countries. In the long-term, global food supplies must increase enough to feed more than 8 billion people; and a fair trading system is vital.

Table 9:1: *Matrix showing the dimensions and critical variables of food security*

	Short term (1-3 years)	Medium-term (5-15 years)	Long-term (25-30 years
HOUSEHOLD	**Access to Food** Nutrition and Health	**Access to Income** Improved Well-Being	**Improved Incomes** Social Infrastructure
NATIONAL	**Safety Nets** Nutrition And Health	**Economic Development** Rural Profitability Agricultural Research	**Food Security For All** Poverty Eliminated
GLOBAL	**Grain Stocks** Food Aid	**International Research** Fair Trading System	Adequate Global Supplies

Source: World Bank (1997: 4).

Although it is difficult to argue against such a vision in its face value, methodologically, the separation between short-term strategies aimed at access to food and medium and long-term policies aimed at access to income and improved income miss the point in separating these inseparable dimensions. However, the matrix is mindful of the complexity within which multi-layered household food security should be conceived.

Fourth, understandably, household food security policies are more difficult for policy makers to deal with than national and global policy issues concerned with the quantity and quality of large economies than small disfranchised small-scale producers. More difficult are issues related to changing households attitudes, consumption patterns and lifestyle. To some extent, this could explain why national and global food security policy and research have been a) the focus of financial allocations and research. It also explains why food security issues b) continued to occupy the minds of policy makers with vested interest in promoting national and global food markets, sometimes even with good intentions, at the expense of household food security.

Fifth, the conversion of the socioeconomic impacts of food production, consumption patterns and lifestyle into environmental change is manifested in soil erosion, deforestation, biodiversity loss, and watershed reclamation for the expansion of food and cash production. In societies where technology, and where land and labor productivity has not kept pace with population, carrying capacity analysts cannot be absolutely right nor absolutely wrong either. Given the current state of agricultural technology in most developing countries, changes in lifestyles and consumption patterns are indicatives of a potential pressure on the environmental and possibly environmental degradation. Climate change in such unfavorable circumstances will not make life any easier.

HOUSEHOLD VULNERABILITY TO FOOD INSECURITY

Food security commonly refers to "access by all at all times to enough food for an active, healthy life. Its essential elements are the availability of food and the ability to acquire it. Conversely food insecurity is the lack of access to sufficient food and can be either chronic or transitory. Chronic food insecurity is a continuously inadequate diet resulting from the lack of resources to produce or acquire food. Transitory food insecurity, however, is a temporary decline in a household's access to enough food. It results from instability in food production and prices or in household incomes. The worst transitory food insecurity is famine".[4] Household food security can be summarised into four interrelated aspects: availability, access, stability and quality.[5]

The factors that contribute to or undermine household food security are manifold. To be sure, an expanded definition of household food security encompasses more than food production. It is linked to the wider environment (physical, socio-economic, political and policy) within which the households eke their living. In rural Africa as in many other developing countries, households capacity to secure food security is often influenced a host of factors such assets, income, entitlement as well as capacity to diversify food sources through purchased, cultivation or collection of or wild edible species.

Defining household food security and the factors that influence it in those terms would challenge the plausibility of the conventional components of food policy (e.g. buffer stock systems, food price stabilisation programmes, food aid, financial food facility schemes, trade policies and food imports mix). In fact, each national food security policy component has its merits and demerits, while some may work under certain macro-economic policy environment, others may be flouted by factors not counted for during the early stage of policy development.

I argue elsewhere that although food policies are important instruments in fostering food security, these are often influenced by other non-food related policies and are restricted by a host of internal and external constraints pertaining to the complex nature of the interplay between food production, consumption and distribution. Therefore food security policies and strategies alone cannot solve the problems associated with food entitlement nor the macro-economic orientations of states in crisis. In the African context, food security policies have provided poor answers to urgent needs, not only for delivering instant solvency to chronic food crisis or heeded the quest for proper reallocation of resources to achieve long-term food security objectives.[6]

While national food security policies are often designed under the assumption that food self-sufficiency is a prerequisite to food-reliance that can hitherto be translated into household food security (ref. to previous section of this paper on World Bank 1997). Experience has shown now and again that household food security can decline even when laudable national food policies are in place. The implication here is that, in most developing countries there is a widening gulf between the perception and reality of national food security policy objectives and household food security determinants.[7] Conversely, countries that are not self-sufficient or self-reliant have been able to deliver a semblance of food security to disfranchised impoverished populations.[8]

Household vulnerability to food insecurity can be defined within a generalised notion of vulnerability. It denotes that a household's can be described as vulnerable to food insecurity when it is susceptible to food inaccessibility, unavailability and instability and/or inadequate quality food. To be sure a household can be vulnerable to food insecurity when the household's capacity to anticipate, cope with, resist and recover from transitory food insecurity.[9] Implicit in this definition is the fact that the magnitude of household vulnerability to food insecurity can be transitory or chronic. In other words, vulnerability cannot be treated as a static condition, but a condition that can be influenced, as food production is, by internal and external factors some of them are beyond the capacity of households to anticipate. These include global climate change, severe periodic or seasonal droughts, epidemics, sudden collapse of commodity markets, change of government policy orientation due to internal or external pressures etc. The factors that may influence the food requirements and food situation of households are also factors that in my view are destined to increase the household vulnerability.[10] These are as follows:

- Household size and composition;
- Access to land;
- Enabling physical environment;
- Time availability;
- Knowledge of appropriate food processing and storage techniques;
- Income;
- Market access;
- Expenditure pattern;
- Pricing policies;
- Power structure;
- Infrastructure and institutional capacity

The significance and magnitude of each factor in augmenting or undermining food security differs household food security differ from household to household. For instance, while some households may have sufficient land, the household size and composition may be inadequate. In other instances the problem could emanate from lack of appropriate knowledge of food storage and preservation hence inability to create buffer stocks.

Household vulnerability to food insecurity occurs as a result of single or multiple failures of production, institutions, markets and the physical environment. In agrarian or livestock producing households the production, market and institutional arrangements are interlocked in and dependent on the physical environment, which provides the life support system necessary for household production system's maintenance. Likewise, market failures or fragmented markets and weak institutions (lack of price surveillance) may deprive households of a substantial proportion of their production (huge surplus appropriation, inadequate market information system, lack of access to markets and powerful middlemen). While the sub-system failure can develop short term survival strategies, regular occurrence of partial or total system failures, as in the case of famine, develop transitory household food insecurity into chronic food insecurity i.e. famine.

Households that draw much of their income from off-farm activities such as petty trade, construction, carpeting, tailoring, domestic labour for well-to-do households, or recipients of regular remittances from migrant members of the household (i.e. can generate an income independent of climatic variability) are better off in withstanding agricultural or pastoral system failures.

In the following I report two of my research experiences. The first is on food production systems in northern Nigeria shows that, although the availability and quality of land, food policy environment largely influence production, institutional weakness and market failure rather than household size and composition patterns are among the main determinants of food insecurity.[11] This finding is particularly relevant for households that have better access to relatively modern technology (plough, fertiliser, knowledge of food storage and preservation) and awareness of pricing policies through a regular flow of market information.[12]

The second is in Tigray region of Northern Ethiopia reveal that despite a commendable intervention by several donors, including the European Union Integrated Food Security Project, the possibility of attaining sustainable household food security is most difficult both at the medium and long-term. Climatic unpredictability, vulnerability to drought, inadequate natural resource base, land fragmentation (less than 2 acres per household in some area), poor technology create an almost impossible condition for long-term household food security.

Using the strict definition of food security (access, availability, stability and quality) as a parameter to determine which factors are more prominent in enhancing food insecurity and household viability is difficult to verify unless large-scale representative sample is used. The field studies that I conducted in Sudan, Nigeria and Ethiopia have convinced me that there is need for comparative method to verify the validity of many of the assumptions that we have held supreme in debate about food security and household vulnerability.

The current developments reveal unstoppable global market integration in which food accessibility rather than availability are more significant than quality and stability. This finding, however, is neither new nor surprising. It only gives further evidence to the plausibility of the entitlement thesis as lauded by the 1998 Nobel Prize Winner Amartya Sen.[13] The problem here is simply that market integration without alternative sources of income convertible into food would reinforce rather than improve households food security. In fact it will subject household food security to external vulnerability over which the household has no control.

Although entitlement failure argument assigns more weight to the social and political analysis rather than environmental or climatic processes, it is now generally recognised that the relationship between household vulnerability to food insecurity and environmental degradation is a miniature of more severe conditions of cumulative long-term climatic variability. An irreversible alteration of the global climatic conditions (inferring from a worst scenario) will lead also to permanency in food shortages. In other words, global climatic change with the magnitude described by some climatologists can potentially transform transitory food insecurity into chronic or perpetual food insecurity. Some elements of food security, particularly those, which are climate (environmental) dependent, would aggravate household vulnerability. Although global climate change could be irreversible, then its consequences require long-term solutions and not short-term remedies.

Three problems are associated with our conventional understanding of household food security: 1) It seems that attempt to integrate global meteorological modelling, local environmental change and household food security is still conceived within in the conventional divide between the social and natural world. 2) As a consequence of the above, the polarisation of research into environmental or socio-economic determinants of food insecurity has hampered efforts to go beyond the obvious linkages between the degradation of environmental life support system and food production. 3) Separation rather integration has dominated the research scene thus ushering in fatalism nurtured by a grim picture of a world melting away, while a helpless humanity is gazing at an incoming doomsday. In this case privileging an abstract global has continued to forsake locality and community in favour of larger synthesis little understood by the local communities, although it impacts on them disastrously.

The impact of global climate change on food production can be altered only if alternative food production systems and coping strategies can be found. This topic is larger than what this presentation intends to achieve and all what I intend to do is to summarise some the indicators to the emerging household coping strategies. In the following section climate change enters the debate on household food insecurity from at least three food production sub-sectors: horticulture, wild edible food and watershed cultivation.

EXPLORING THE LINKAGES: HOUSEHOLDS COPING WITH CLIMATE CHANGE

The relationship between the production of staple food and cash crops and climate change particularly in the case of prolonged drought is very well documented. In relation to hazards, Blaikie et al (1993:61-70) reviewed and summarised the literature on household coping strategies according to the following:[14]

- Preventive strategies;
- Impact minimising strategies;
- Creation and maintenance of labour power;
- Building up stores of food and saleable assets;
- Diversification of the production strategy
- Diversification of income sources;

- Development of social support networks

These household coping strategies deal with short term to medium term events climate change events seen with a wider perspective of local resource management and allocation in order to offset food security. Therefore we have to be very cautious in attributing them to long-term cumulative processes such global climate change. At least three of the conventional strategies such preventive, building up stores of food and diversification of production are climate dependent.[15] Furthermore, these strategies cannot be realised without substantial improvements in water management, soil conservation as well as the introduction of new technologies including drought resistant crops.

Although the above strategies surface in the literature, with varying degree of emphasis on one strategy or the other depending again on the general health of the environment, the societal responses to success of failure are well documented.[16] However, less documented are the consequences of drought and other climatic variables on horticulture, wild edible plant species and watershed cultivation, that constitute the main sources of food during drought and impaired climatic functions.

Wild edible plants: National and global food policy circles have not seriously acknowledged the contribution of wild edible plants to household food security. This could be due to the common perception that food consists only of domesticated edible plants and animals or that food policies are informed by elite who have long lost interaction with the agrarian and pastoral societies from which the have originated.[17] However, among the few works that I would like to bring to the attention of the reader is that of EL Tom (1987) which offers an excellent documentation of the use of wild edible plants during famine periods in Dar Fur region of the Sudan.[18]

Wild edible plants are not only an important source of food, but also some of them constitute an important source of medicine as well. It will be pretentious to introduce this section of the presentation, as if it intends to fill this important gap in the literature. However, what I attempt to do here is to present some part of the material which I have collected on the variety of edible plants during my research in the Nuba Mountains of the Sudan (table 9.2) and the Fulani of Northern Nigeria (table 9.3).

The Moro-Nuba peoples inhabit the central part of the Nuba Mountains of the Sudan. Located between Lat. 10:45 and 11:10 degrees North and Lang. 29:45 and 30:15 degrees West, Moroland's climate provides suitable conditions for arable cultivation, animal husbandry, hunting and gathering. The rainy season extends from late May early June to mid October and in rare seasons to early November. The annual rainfall ranges between 450 and 800 mm. The rain increases from North to South where the southern parts receive more rain than the northern fringes boarding the dry Savannah. Although sorghum, maize and a variety of beans and vegetables constitute their main sources of staple food, in recent years, the Moro have become increasingly dependent on the collection of wild edible plants. Two factors have contributed to this: recurrent rain failure (drought) and war and political instability. However, since I have observed the increase in the use of wild edible plants at least five years before the region became part of the war-zone in 1988, it is evident that the 1973-1975 and 1983-1985 droughts have contributed to this trend. Table 9.2, shows some of the edible plants collected by the Moro in order to offset seasonal food deficit (May – August) as well as transitory and chronic food deficits.

Second the case of the Fulani pastoralists of Northern Nigeria. I carried out research in the area between Lat and Long. The droughts of the 197s, 1980s and 1990s have pushed the Fulani from Sokoto State to the Northern frontiers of Illorin State i.e. from a territory of less than 400 mm per annum to an area of an annual rainfall of 800 to1200 mm. Milk and millet constitute the main staple food, although Fulani food habits have changed considerably during the last two decades or so. More Fulani eat imported food (rice and wheat) and although vegetables were not appreciated in the past, onions and a variety of local beans have gradually been appreciated.[19]

Table 9.2: *A sample of edible plants collected by the Moro: Parts eaten, preparation and season*

Nuba/Moro	Latin or English	Edible parts	Preparation	Season
Ngora-gorina	*Portulaca Cloracea*	Leaves	Relish eaten with porridge	June-November
Ngware	*Andonsonia Digitata*	Leaves Fruits	Raw or mixed with sesame Raw or mixed with milk	April-July December-March
Orlo	*Terinalia Laxoflora*	Leaves Fruits	Mixed with sesame Soup or thick relish eaten with porridge	May-August
Angobongar	*Wild fig or Maba abassinica*	Fruits	Raw	December-March
Erdi-uri	*Mushrooms*	Whole plant	Boiled with grain	August-December
Avortinga	*Wild grape*	Leaves and fruits	Relish, eaten with porridge	July-November
Daraba	*Hibiscus esculentus*	Vegetable	Cocked with meat and used as a relish, eaten with porridge	October-November
Ngoi	*Albizzia Amara*	Fruits	Raw	November-December
Haluuk	Palm	Root	Roasted	February-April
Barlo	Corchorus	Leaves	Relish eaten with porridge	June-November
Orio	Wild spiked cucumber	Vegetable	Raw	October-November
Nabag	*Acacia Mauritanica*	Husk and seeds	Raw	Decmber-February

Source: Author's field notes.

During drought Fulani pastoralists use two strategies to reduce dependence on staple food: first, they reduce food consumption by reducing the ration of millet mixed with sour milk (locally called *nano*) and an active search for wild edible plants over 50 percent of their food intake from wild edible plants often scarcely used during good rainy years. Table 9.3 below shows a sample of the wild edible plants collected by the Fulani.

Wetlands cultivation: The official global statistics on wetlands often refers to large-scale freshwater wetlands, mangroves and swamps. Wetlands serve a variety of functions: they assist flood control, purify water supplies, protect shorelines from erosion, and trap sediments that can pollute waterways. Wetlands constitute an important source of water, food (fish, birds, crop), pasture, wildlife, forest and forest products etc. Policies concerned with wetland conservation have concentrated on conventional threats such as alteration of and disruption of natural processes. These processes include agricultural intensification, urbanisation, pollution, dam construction, regional water transfers construction, agricultural expansion, water reclamation or rangelands degradation.

Table 9.3: *A sample of edible plants collected by the Fulani: Parts eaten, preparation and season*

Fulani	Latin or English	Edible parts	Preparation	Season
Barkehi	*Dearuim microcarpum and d. senegalense*	Leaves	Boiled	November
Bummehi	*T. glancescens and Terinalia Laxoflora*	Fruits	Raw	December-March
Butuye	*Kookiya strychnos Spinoso*	Fruits	Raw	October-December
Chambulli	*Ximenia Americana*	Fruits	Raw	December-March
Kahi	*Latin name not known*	Barks	Raw	August-December
Dingali	*Garenia Sokotensis and Gardenia squalla*	Fruits	Raw	November
Ka'rehi	*Butyospermum paradoxum*	Seeds	Oil	October-November
Knokehi	*Dichrostachys cinerea and Dichrostachys dioscorea*	Fruits	Raw	November-December
Narehi	Parkia clappertoniana	Flowers	Roasted	September
Pa'aohi	*Latin name not known*	Fruits	Raw	November
Tane	Balanites Aegyptiaca and Balanites cayjanuscajan	• Fruits • Seeds	• Raw • Oil	September-November
Kuriahii	*Bombax Baunoezence*	Flowers	Raw	August

Source: Author's field notes

African wetlands have been accorded some attention at the international level, but the interest as I have mentioned earlier is on wetlands with potential global consequences on biodiversity and climate change that some local wetlands. [20] The testimony to this was the recognition by the Second International Conference held in Dakar Senegal (1998) of the neglect of African wetlands and the need to empower local communities to manage them better.[21]. A further testimony to the neglect of local small-scale wetlands in the international agenda is the excellent report on African drylands prepared from wetlands International by Stevenson and Frazier (1999). This report shows the meagreness of information on African wetlands at the national level let alone the local level.[22]

Some scientists even infer that large-scale wetland reclamation (or drainage) and destruction of vegetations may accelerate global climate change and turn potentially useful ecosystems into wastelands. However national and global policies dealing with climate change have given little to no attention to local level wetland reclamation by individual households. In my view, the reclamation of wetlands in drylands may as well pose an unnoticed cumulative climatic change parallel to the consequences of large-scale global change.

The evidence I have is not conclusive at the global level, but highly significant for regions with similar climatic conditions i.e. characterised by decreased precipitation, variability and vulnerability to drought coupled with an accelerated wetlands reclamation for cultivation. For instance, the material collected from the Somali pastoralists between Jijiga and Dagahbor in Northeaster Ethiopia reveal that out of 23 large natural water pools (sing. *birka*) only seven pools still hold water some water, which lasts for about three months after the end of the rainy season. In other words, all 23 water pools dry-up by February leaving human and animal populations dependent for water on bore holes. There has also been significant population movement to low depressions and valleys where they can have better pasture and more fertile lands.

In Northern Nigeria, my research local Hausa farmers and Fulani pastoralists in the area from Okuta (in Illorin State, Nigeria) to the borders with Benin reveal similar pattern. None of the 67 large water pools that were frequented by Fulani pastorsalists in the Northwestern parts of Borgu District of Illorin State, were still existent. Similar to the case of North-east Ethiopia, low depressions and water pools became a source of fertile land, more moisture that can keep long and offers a more predictable crop relative to drylands cultivation.

In my view local level small-scale wetland reclamation in arid and semi-arid lands climate is one of the household's main coping strategies during deteriorating precipitation conditions. However, once the households have settled and managed to reap better incomes and food security from a more predictable resource base in the wetland, their settlement becomes permanent.

However, it is important to note that drylands reclamation is often accompanied by social and political tensions emanating from their scarcity and subsequent competition for their ownership. A number of recent studies, not specifically on wetlands, have shown that land scarcity due to land concentration can contribute to conflicts of different magnitude.[23]

Another consequence of local wetland reclamation is biodiversity loss. No serious studies or simple recording of the biodiversity resources of regions were conducted. Remote areas, with valleys, swamps or watersheds essential for recharging large watershed often do not occupy a high position in global interests in the large and more visible.

The question that would continue to haunt conservation and development advocates alike is whether it is possible for households and local communities to use the wetlands for the twin purposes of conservation and food security. In view of this, precipitation and climate change are more important than we have thought. They are not merely about change in the weather or global strategies and legislations for advancing authoritarian conservation. They are matters concerned with empowering local communities and households in order to be able to negotiate better strategies to reverse or reduce the impacts of climate change on the socio-economic and environmental conditions available to them.

Horticulture: Generally horticulture refers to the growing of fruit (especially tree fruits or pomology), production of vegetable crops (olericulture); production of flowers (floriculture), and ornamental horticulture (landscape gardening). In this section I refer specifically to the three former varieties of horticulture, which are attracting increasing attention in Africa.[24]

By nature horticulture is more predictable than agriculture, often combines the use of seasonal rainfall as well as more permanent sources of water (large or small-scale, modern or traditional irrigation systems). The possibility of diversifying water sources reduces horticulture's vulnerability to climatic variability. More over, it is argued that horticultural products, particularly fruits and vegetables fetch higher prices, have high nutrition value and could therefore improve not only the incomes of the households engage in it but also offer better nutritional values.

Although my research on various aspects related to the contribution of horticulture to household food security is not complete, I noted that among the Moro of the Nuba Mountains traditional horticulture plays an important role in the household food security. The Moro divide their farms into three types: 1) *Jubraka* (or house gardens) for growing early maturing crops. 2) *Hadaba* (light soils plateau) and *fawa* (black cotton soil) for growing late maturing grain and cash crops such as ground nuts, sesame and cotton. 3) *Jinaina* (literally garden a small plot on the low depressions or water courses before they completely dry-up (similar to flush cultivation) for growing maize, pumpkins, tomatoes, cucumber and other minor crops.

Table 9.4: *Vegetables and fruit production in 1000 kg 1983 and 1993*

Vegetables and Fruits	Total production (1983)	Total production (1993)	% increase 1983/1993
Sweet Potatoes	12.7	42.2	332.28
Hybrid Tomatoes	4.9	13.7	279.59
Aubergines	3.7	6.6	178.37
Paprika	1.5	5.2	346.66
Onion	9.3	35.8	384.94
Fruits			
Mangoes	2.6	12.8	492.30
Guava	1.7	4.6	270.58
Grape fruits	3.7	4.7	127.02
Orange	2.5	4.4	176
Citron	2.3	8.7	378.26

Source: Author's field notes

During the last decade or so an increasing number of farming communities began to establish orchards for the growing of citrus fruits (citron, orange and grapefruits), mangoes, guava and others for local consumption and export. New vegetables and fruits have also been introduced into the area such as spring onions, hybrid tomatoes, aubergines and watermelons. Table 9.4 shows the types of horticultural crops introduced into the area and the quantities produced, comparing 1983 and 1993. The increase in fruits and vegetable production can be explained among other things against the increase in the number of household producers from 21 in 1983 to 87 in 1993 (i.e. over 4 times). However due to the fact that not all farmers grew more varieties of fruits or vegetables, the increase in production is disproportionate with the increase in the number of producers. What the table does not show is the high wealth differentiation among fruit and vegetable (cash, livestock), access to labour (household size), access to fertile lands within proximity from water courses, distance from rural and urban markets etc. What is important here is that climate change and other factors related to changes in lifestyle and consumption patterns have contributed to diversification of production strategy including the introduction of new crops, technology and other modern input delivery systems.

Hence, the three household coping strategies (use of wild edible plants, wetland reclamation and horticultural production) that I attempted to explain are not exclusive or unrelated. Horticultural production takes place in low depressions and fertile wetlands reclaimed by households that foresaw their capacity to produce cash crops that are highly demanded by the market in order to generate higher incomes.

However, households' wetlands reclamation contribute to biodiversity loss and deprives the whole community important sources of food, particularly mud fish and a large variety of wild edible plants as explained in tables (9.2 and 9.3. While conventional household food security studies have concentrated on the interplay between population and resources, the new household food security agenda should be mindful of the long-term environmental consequences of these strategies. To that extent, conventional household food security can be embedded in a zero sum game where long-term dependence on wetlands and wild edible plants the sustainability of the production system as well as in-situ biodiversity prospects.

CONCLUSION

Climate change poses a serious challenge to real societies and communities and cannot be explained in terms of abstract global notions removed from these realities. In this chapter I tried to explore the linkages between climate change, wetlands reclamation, biodiversity and the impact of that on local communities' ability to eke their living from a productive environment. Dialectically, climate change will force communities to reclaim the wetlands for more reliable sources of food; on the other such source of food security would undermine wetlands conservation and contributes to biodiversity loss.

No matter whether climate change is real, imagined, due to human-made or natural, the scientific-social science debate and policy interventions should be aware that the current levels of climatic variability is a grim sign of what is lying ahead. The severity and potential frequency of extreme climatic conditions is difficult to predict, but recent history has shown that even sporadic extreme conditions are difficult to cope with. What is important here is to realise that global processes contributing to increasing greenhouse gases (GHG) concentration and communities and households feel extreme events more directly locally. If these local communities and households are ill prepared to cope with the current drought frequency, it will be too hard to imagine how they would cope with extreme climatic conditions.

The case studies introduced in this chapter illustrate that as the intensity and frequency of climate change and variability increase, households tend to develop coping strategies and adapt to situations different from those of the past. Resources that were previously under-exploited or resources that were not fully utilised such as wetlands are increasingly exploited. It is natural that households under food security stress look out for more predictable production system even if this meant undermining the resource base on which they survive. The combined effects of extreme climatic conditions and overexploited wetlands would leave the local communities with meagre to no alternative sources of livelihood. Less however, is known about the pattern and magnitude of biodiversity loss in localised small wetlands, let alone the variety and quality of wild edible plants and other species not yet counted for? The rate of reclaimed wetlands degradation in hydrologic systems that are not of international significance has hardly been documented least its long term consequences of household vulnerability to food insecurity been documented.

At issue here is not global climate change studied in isolation from sustainable agriculture, the latter in isolation from sustainable household food security, socio-economic conditions or natural processes vital for enhancing or eroding the generative capacity of the environment in isolation from the rest. What is at issue here is the discernible linkage that we have explored between sustainable agriculture that depends on the reclamation of wetlands and the long-term future sustainability of both the environment and household food security. Unless local communities and households food security is treated within the wider context of other environmental functions, global policies on climate change, biodiversity, wetlands of international significance and food security may find themselves in a collision course. As has often been the case, in some instances, international institutional politics could be more damaging to local livelihoods than controllable natural disasters. Hence, even if all socio-economic aspects of food security (entitlement, security of tenure, rural credit, technological transformation, favourable macro-economic policies, decentralisation and local participation etc.), were solved, the long-term sustainability of these will in jeopardy.

The implication of the preliminary findings of this chapter for research on the twin objective of sustaining the regulative functions of the environment while maintaining household food security by reducing vulnerability to climate variability is daunting. The chapter also alerts us to the fact that climate change research has been suffocated by "scientific" polemic about its eminence or remote possibility. Or that what is required is the raising down of barriers between in the bid to assign prominence to peoples' futures over whether climate change is real or imagined. The key answer to this question could be is to go back to basics in a genuine search for methodologies capable of recognizing the unity of the social and natural world particularly for policy research where the integrative nature of the environmental functions become inseparable of the socio-economic processes through which livelihood security, in all its manifestations is produced and reproduced. Such an approach may as well provide an opportunity for a better synergy, coordination and intergaration of global environmental policy regimes.

ENDNOTES

1. See Okoth-Ogendo, H. W. and J. B. Ojwang (Eds.) (1995) *A Climate for Development: Climate Change Policy Options for Africa*. Nairobi: African Centre for Technology Studies.
2. Reference here is made to the increasing frequency of Al Nino phenomenon and climate unpredictability. See for instance, Michael H. Glantz (2001) *Currents of Change: Impacts of El Nino and LA Nina on Climate and Society*. Cambridge University Press.
3. For more material on the food industry and food security see Hulse, J. H. Science, Agriculture and Food security. Ottawa, Ontario: Nation Research Council press. Bonanno, A. (1994) from Columbus to Conagra the Globalization of agriculture and food. Lawrence, KS: University Press of Kansas. Bernstein, H., Crow B. Mackintoch, M, and C. Martin (1990) the Food Question: profits versus people. London: Earthscan Publications. Watkins, K (1995) Agriculture, Trade and Food Security. London: Oxfam.
4. Reutlinger, S. (1987) "food security and poverty in developing countries", in j. P. Grittinger J. Lislie and C. Hoisington (Eds.) *Food Policy: Integrating Supply, Distribution and Consumption*, the John Hopkins University Press, Baltimore and London.
5. See Groverman, V. (1997) *Strategies for Food and Nutrition Security at the Household Level*, Ministry of Foreign Affairs, The Hague and Smith, M. (1993) *Household Food Security: Concepts and Definitions: An Annotated Bibliography*. Institute of Development Studies (IDS), Brighton.
6. Mohamed Salih, M. A. (Ed.) (1992) *Inducing Food Insecurity: Perspectives on Food Policies in Eastern and Southern Africa*, Scandinavian Institute Of African Studies, Uppsala.
7. Davies, S And M. Leach (1991) Globalism Versus Villagism: National and International Issues in Food Security And The Environment, in IDS Bulletin, Vol. 22, No. 3, Pp. 43-49. Institute of Development Studies, Sussex.
8. See Hollist, W. L. And Tullis, F. Lamond (Eds) (1987) pursuing food security: strategies and obstacles in Africa, Asia, Latin America and the Middle East. Boulder and london: lynne rienner publishers.
9. This definition is compiled with modification from three sources: 1) Lewis, James (1999) *Development In Disaster-Prone Places*: Studies in Vulnerability (1999) Intermediate Technology Publications. 2) Ribot Jesse Craig, Antonio R. Magalhaes, Stahis Panagides (Editor) (1995) *Climate Variability, Climate Change And Social Vulnerability in the Semi-Arid Tropics*, Cambridge University Press 3) Blaikie, P., T. Cannon, J. Davis and B. Wiser (1994) *At Risk: Natural Hazards, Peoples Vulnerability and Disasters*. Routledge, London and New York.
10. See Groverman, V. (1997) *Strategies for Food and Nutrition Security at the Household Level*, Ministry of Foreign Affairs for more details.
11. I carried out research in Borgu district, Northern Nigeria and the Nuba Mountains, Sudan. The figure above is a generalised model of food production in an agro-pastoral production system embedded in a physical environment and

interacting with markets and institutions. The figure is not fully explain nor its historical dimension(s) colored in.

12. Mohamed Salih, M. A. (1987) *Agrarian Change In The Central Rainlands, Sudan*, Uppsala: Scandinavian Institute of African Studies. Mohamed Salih, M. A. (1992) "Environmental and Social Insecurity in The Drylands of The Sudan", In Hjort, af Ornas (Ed) *Security in African Drylands*: Research, Development and Policy, Uppsala: Department of Human And Physical Geography. Mohamed Salih, M. A. and D. Eshete (1997) *Evaluation of Novib/European Union Integrated Food Security Project in Tigray Region*, Ethiopia. The Hague: Institute of Social Studies.
13. Sen, Amartya (1983) *Poverty and Famines*, Oxford University Press And Sen, Amartya (1999) *Development as Freedom*, Knopf Alfred A. Particularly Chapter 4 On "Poverty as Capability Deprivation", Pp. 87-110 and Chapter 7 on "Famines and Other Crises", pp. 160-188.
14. Blaikie, P. T. Cannon, I. Davis and B. Wisner (1993) *At Risk: Natural Hazards, Peoples Vulnerability and Disasters*. London And New York: Routledge. For The Particular Case of Climate Change and Africa See Ominde, S.H. and C. Juma (Eds.) *A Change on the Weather: African Perspectives on Climatic Change*. Nairobi African Centre For Technology Studies (Acts) Pesss.
15. For a more relevant case to the sensitivity of household food security to climate change see Downing, Thomas and M. L. Parry (1991) Climate Change, Agriculture And Food Security, In Ominde and C. Juma (Eds.) *A Change In The Weather: African Perspectives on Climatic Change*. Nairobi African Centre For Technology Studies (ACTS) Pesss.
16. El Tom, A. O. (1987) "Implications Of The 1983-1985 Famine On The Rural Sahel Of Sudan", In M. A. Mohamed Salih (Ed.) *Agrarian Change in the Central Lands of the Sudan*. Uppsala: Scandinavian Institute of African Studies.
17. Research on wild edible plants in Africa is most retarded. The only existing serious work is Goode, P. M. (1989) *Edible Plants of Uganda: The Value of Wild and Cultivated Plants as Food*. Food and Agriculture Organization of the United Nations. Although most of the Literature on wild edible plants comes From the United States of America And India, One of the most world-wide cited references on this subject Is Sholto Douglas, James (1978) *Alternative Foods: A World Guide to Lesser-Known Edible Plants*. London and Pelham: Levittown, N.Y.
18. El Tom's (1987:152) material collected during the 1983-1985 famine in Dar Fur region of the Sudan also confirms this view. El Tom mentions four wild edible species:
19. The most popular edible wild plant is a wild berry called *mukheit* (*boscia senegalesis*), this is the fruit of a semi-desert tree, which flourishes in times of drought; in fact, its presence is regarded as an indicative of plant failure. *Mukheit* is popular because it is easy to collect, and in summer the fruits are obtainable at the peak of food shortage just before the first showers. *Mukheit* is however, poisonous and is only edible after careful and tedious processing. It is dried and then soaked in water for at least 12 days, during which it is washed several times. This processing also intended to eliminate the bitter taste of the fruit.

20. The seeds of koreib (*barachiari or panicum laetum*?), A semi desert annual plant are collected during season, long before millet ripens, also a period of acute food shortage. The seeds are collected dried and latter ground and used as a substitute for millet.
21. *Haskanit (*Cenchrus Biflorus) is a wild thorny plant also used as a substitute for millet. It retains its seeds much longer than *Koreib* and therefore kept as a last resort.
22. For more material refer to M. A. Mohamed salih (1992) *Pastoralists and Planners: Local Knowledge And Resource Management In Gidan Magajia Grazing Reserve, Northern Nigeria*. International Institute of Environment and Development, Policy Paper No. 32.
23. As of November 1999, there are 116 signatories of the Ramsar Convention of Wetlands of International Significance (adopted 2nd february1971 in Ramsar, Iran). 27 African countries are among the signatories. African-Eurasian Migratory Water-birds Agreement (AEWA), the largest of its kind developed so far under CMS, was concluded on 16 June 1995, when representatives of over 66 countries met in The Hague, Netherlands. The AEWA covers 172 species of birds ecologically dependent on wetlands for at least part of their annual cycle, including many species of pelicans, storks, flamingos, swans, geese, ducks and waders. The agreement concerns 117 countries (plus the European Union) from Europe, Parts of Asia and North America, the Middle East and Africa. In fact, the geographic area covered by the AEWA stretches from the northern reaches of Canada and the Russian federation to the southernmost tip of Africa. Various projects are already envisaged under this agreement, among them: promotion of biodiversity conservation in wetlands, rehabilitation of former brown-coal pits in Germany, monitoring of bird migration in the middle east, and habitat conservation in African countries. Various species conservation plans have also been prepared. Most of them will serve as models to facilitate implementation of the agreement after its entry in force.
24. African wetlands offer exciting opportunities for sustainable development: they are frequently characterised by direct sustainable use of wetland products by local communities; they often incorporate highly sophisticated traditional management systems; and they are increasingly the focus of flourishing eco-tourism initiatives. There are many lessons here for other regions of the world. Yet African wetlands have received inadequate attention. For instance, large-scale development projects have often had catastrophic impacts on wetland functioning; national agencies responsible for wetland conservation are severely constrained by resource limitations and a lack of trained personnel. This limited capacity hinders the attraction and management of large-scale bilateral and multilateral funds for wetland conservation. Local realities make it hard to implement national legislation let alone international conventions and tereaties.
25. Stevenson, N. and Frazier, S. (1999) *Review of Wetland Inventory Information in Africa, Wetlands*. Wageningen: International–Africa, Europe, Middle East.
26. Among recent publications on this topic Markakis, J (1998) *Resource Conflicts in the Horn of Africa*. London: Sage publications. Mohamed Salih, M. A. (1999) *Environmental Politics and Liberation in Contemporary Africa*. Dordrecht. London And New York.

27. In this section a distinction is made between agriculture and horticulture whereby agriculture refers to more open culture of agriculture such as the production of grains and grasses, known as agronomic crops, which are cultivated on a large scale. The original distinctions have been so blurred that many crops formerly considered either agronomic or horticultural are now categorised sometimes in one field, sometimes in the other, depending on the intended use of the crop. Thus a plant grown for home consumption may be called horticultural; the same plant cultivated for forage is regarded as an agronomic crop (adapted from S. Pickett, for former Head of the Department of Horticulture, University of Tennessee and the Tennessee Agricultural Experiment Station). In the particular case of Africa see for instance, Mahmoud S. Hassan and H. Burgstaller (1983) *Horticultural Crops: Horticultural Research and Development.* 8th Africa Symposium on Horticultural Crops 20-24 March, Wad Medani. Tjalling, Dijkstra (1997) 'Commercial Horticulture By African Smallholders: A Success Story from the Highlands of Kenya'. *Scandinavian-Journal of Development-Alternatives.* 16, 1, Mar, 49-74.

Notes on Contributors

Edlam Aberra, was Assistant Lecturer at the Department of Sociology and Social Work, College of Social Sciences, Addis Ababa University, Ethiopia. Currently she is a postgraduate student at Cambridge University, UK.

Abdel Ghaffar Mohamed Ahmed is Professor of Development Anthropology. Currently he is the Executive Secretary of the Organisation for Social Science Research in Eastern and Southern Africa (OSSREA), Addis Ababa, Ethiopia. His Publications include *Sustainable Development in Eastern and Southern Africa* (Macmillan and St. Martins Press 1998) and *African Pastoralism: Conflict Institutions and Government* (Pluto Press 2001)

Zie Gariyo is Co-ordinator of Debt Network, a Non-Governmental Organisation (NGO) engaged in policy analysis and advocacy. He is also Research Fellow at the Centre for Basic Research (CBR), Kampala, Uganda.

Henry Manyire is Lecturer at the Department of Women and Gender Studies and teaches gender and development, economics and methods of social science research at Makerere University, Kampala, Uganda.

Paul S. Maro is Professor of Geography, currently serving as technical advisor with Southern African Development Community Environment and Land Management Sector (SADC-ELMS), Maseru, Lesotho.

Samson Wokabi Mwangi is Lecturer in urban development at the Department of Sociology, Egerton University, Nakuru, Kenya. Currently he is Research Fellow at Amsterdam Research Institute for Global Change (AGIS).

Wilson Nyaoro is Lecturer on environment and resource management at the Department of Environmental Management and Planning, Kenyatta University, Nairobi, Kenya.

Fuuna Peter is Gender officer, working for Africa 2000, and a researcher by profession, specialised on gender focused resource management and sustainable development in Eastern Africa.

M. A. Mohamed Salih is Professor of Politics of Development at the Institute of Social Studies, The Hague and the Department of Political Science University of Leiden, The Netherlands. His books include *Environmental Planning, Policies and Politics in Eastern and Southern Africa* (Macmillan and St. Martin's 1999), *Environmental Politics and Liberation in Contemporary Africa* (Kluwer Academic Publishers 1999) and *African Democracies and African Politics* (Pluto Press 2001).

Index

MIX
Papier aus verantwortungsvollen Quellen
Paper from responsible sources
FSC® C105338

If you have any concerns about our products,
you can contact us on
ProductSafety@springernature.com

In case Publisher is established outside the EU,
the EU authorized representative is:
Springer Nature Customer Service Center GmbH
Europaplatz 3, 69115 Heidelberg, Germany

Printed by Libri Plureos GmbH
in Hamburg, Germany